T0235126

SpringerBriefs in Molecular Science

Biobased Polymers

Series editor

Patrick Navard, CNRS/Mines ParisTech, Sophia Antipolis, France

Published under the auspices of EPNOE*Springerbriefs in Biobased polymers covers all aspects of biobased polymer science, from the basis of this field starting from the living species in which they are synthetized (such as genetics, agronomy, plant biology) to the many applications they are used in (such as food, feed, engineering, construction, health, ...) through to isolation and characterization, biosynthesis, biodegradation, chemical modifications, physical, chemical, mechanical and structural characterizations or biomimetic applications. All biobased polymers in all application sectors are welcome, either those produced in living species (like polysaccharides, proteins, lignin, ...) or those that are rebuilt by chemists as in the case of many bioplastics.

Under the editorship of Patrick Navard and a panel of experts, the series will include contributions from many of the world's most authoritative biobased polymer scientists and professionals. Readers will gain an understanding of how given biobased polymers are made and what they can be used for. They will also be able to widen their knowledge and find new opportunities due to the multidisciplinary contributions.

This series is aimed at advanced undergraduates, academic and industrial researchers and professionals studying or using biobased polymers. Each brief will bear a general introduction enabling any reader to understand its topic.

*EPNOE The European Polysaccharide Network of Excellence (www.epnoe.eu) is a research and education network connecting academic, research institutions and companies focusing on polysaccharides and polysaccharide-related research and business.

More information about this series at http://www.springer.com/series/15056

Rodolphe Sonnier · Aurélie Taguet
Laurent Ferry · José-Marie Lopez-Cuesta

Towards Bio-based Flame Retardant Polymers

Springer

Rodolphe Sonnier
Centre des Matériaux des Mines d'Alès
Ecole des Mines d'Alès
Alès
France

Laurent Ferry
Centre des Matériaux des Mines d'Alès
Ecole des Mines d'Alès
Alès
France

Aurélie Taguet
Centre des Matériaux des Mines d'Alès
Ecole des Mines d'Alès
Alès
France

José-Marie Lopez-Cuesta
Centre des Matériaux des Mines d'Alès
Ecole des Mines d'Alès
Alès
France

ISSN 2191-5407   ISSN 2191-5415 (electronic)
SpringerBriefs in Molecular Science
ISSN 2510-3407   ISSN 2510-3415 (electronic)
SpringerBriefs in Biobased Polymers
ISBN 978-3-319-67082-9   ISBN 978-3-319-67083-6 (eBook)
DOI 10.1007/978-3-319-67083-6

Library of Congress Control Number: 2017952522

© The Author(s) 2018
This work is subject to copyright. All rights are reserved by the Publisher, whether the whole or part of the material is concerned, specifically the rights of translation, reprinting, reuse of illustrations, recitation, broadcasting, reproduction on microfilms or in any other physical way, and transmission or information storage and retrieval, electronic adaptation, computer software, or by similar or dissimilar methodology now known or hereafter developed.
The use of general descriptive names, registered names, trademarks, service marks, etc. in this publication does not imply, even in the absence of a specific statement, that such names are exempt from the relevant protective laws and regulations and therefore free for general use.
The publisher, the authors and the editors are safe to assume that the advice and information in this book are believed to be true and accurate at the date of publication. Neither the publisher nor the authors or the editors give a warranty, express or implied, with respect to the material contained herein or for any errors or omissions that may have been made. The publisher remains neutral with regard to jurisdictional claims in published maps and institutional affiliations.

Printed on acid-free paper

This Springer imprint is published by Springer Nature
The registered company is Springer International Publishing AG
The registered company address is: Gewerbestrasse 11, 6330 Cham, Switzerland

Contents

About the Authors

José-Marie Lopez Cuesta received his engineer diploma and Ph.D. degree from Institut National Polytechnique de Grenoble. He is full professor since 2007 and head of the Materials Research Centre of Ecole des Mines d'Alès. He is co-author of around 120 papers in international journals about polymers and composites. He coorganized in 2015 the Eurofillers and Polymer Blends International conference. He is also secretary of the thematic group "fire reaction of organic materials" of the French Chemical Society (SFC). His competences include the study of the microstructure of polymer micro- and nanocomposites, the modification of their interfaces and the development of innovative flame retardant systems.

He has written the first chapter devoted to Flame retardant biobased polymers.

Laurent Ferry is professor in Ecole des Mines d'Alès. For more than 20 years, he has developed research activities in the field of polymer degradation and polymer aging. Since 2009, he is leader of the group working on flame retardancy of polymeric materials in Centre des Matériaux des Mines d'Alès (C2MA). He has worked a lot on the use of micro- and nano-fillers as flame retardant. For several years now, he has been interested in the development of biobased solutions to improve the fire behavior of polymers.

He has written the second chapter devoted to Biobased Flame Retardants.

Rodolphe Sonnier is graduated from ENSAIT (Ecole Nationale Supérieure des Arts et Industries Textiles, France's 1st school of textile engineers) and received his Ph.D. degree in chemical sciences from the University of Montpellier. He joined C2MA in 2008 as an Armines research engineer and later as assistant professor of Ecole des Mines d'Alès. He is co-author of around 70 articles in international journals. His competences include the study of fire behavior and fireproofing of polymer materials, the compatibilization of polymer mixtures and the modification of polymers by ionizing radiation (γ and β).

Aurélie Taguet gained her Ph.D. in 2005 in the field of Polymer Science. In 2008, she became Research Associate in C2MA/Mines-Ales.IMT. She focuses on how morphologies can tailor the final thermal, mechanical and fire properties of bio-composites and nanocomposites? She has already published 30 articles in peer-reviewed international journals and two book chapters (h-index=11). Throughout all her career, she has acquired skills on surface modification of mineral fillers (especially silicates), processing of polymers, polymer blends and composites, and characterizations of polymer based materials especially thermal, mechanical and fire properties.

Aurélie Taguet and Rodolphe Sonnier have written the third chapter devoted to Flame retardancy of natural fibers reinforced composites.

Abbreviations

ABS	Acrylonitrile-butadiene-styrene
APP	Ammonium polyphosphate
ATH	Alumina trihydrate
BAMPO	Bis(m-aminophenyl)methylphosphine oxide
CF	Carbon fiber
CNF	Cellulose nanofiber
CNT	Carbon nanotube
DAP	diammonium phosphate
DDM	Diaminodiphenylmethane
DDS	Diaminodiphenyl sulfone
DGEBA	Diglycidyl ether of bisphenol A
DGPP	Diglycidylphenylphosphate
DNA	Deoxyribonucleic acid
DOPO	9,10-dihydro-9-oxa-10-phosphahenanthrene-10-oxide
ECHA	European Chemical Agency
EDS	Energy dispersive X-ray spectroscopy
EG	Expanded graphite
EVA	Ethylene vinyl acetate
FR	Flame retardant
FTIR	Fourier transform Infrared
GF	Glass fiber
GPCS	Glycidyl methacrylate of chitosan phosphate
HDPE	High density polyethylene
HRR	Heat release rate
IFR	Intumescent flame retardant
IPDA	Isophorone diamine
LbL	Layer by layer
LDH	Layered double hydroxide
LLDPE	Linear low density polyethylene
LOI	Limiting oxygen index

LSH	Lignin-silica hybrid
MAHRE	Maximum of average heat release evolved
MAPC1	Dimethyl(methacryloxy)methyl phosphonate
MAPE	Maleic anhydride-grafted polyethylene
MCAPP	Microencapsulated ammonium polyphosphate
MCC	Microcrystalline cellulose
MDI	Methylene diphenyl diisocyanate
MMT	Montmorillonite
MP	Melamine phosphate
MPCS	Melamine salt of chitosan phosphate
MVP	Dimethylvinylphosphonate
MWCNT	Multi-walled carbon nanotubes
NCC	Nanocrystalline cellulose
NF	Natural fiber
NFC	Nanofibrillated cellulose
NS	Nanosponge
ODPA	Octadecylphosphonic acid
oMMT	Organically modified montmorillonite
PA	Phosphoric acid
PA6	Polyamide 6
PALF	Pineapple leaf fiber
PBS	Polybutylene succinate
PBT	Polybutylene terephtalate
PC	Polycarbonate
PCS	Phosphorylated chitosan
PE	Polyethylene
PEC	Polyelectrolyte complex
PEI	Polyethyleneimine
PER	Pentaerythritol
PET	Polyethylene terephthalate
pHRR	Peak of heat release rate
PLA	Polylactic acid
POSS	Polyhedral Oligomeric Silsesquioxane
PP	Polypropylene
PPA	Polyphosphoric acid
PU	Polyurethane
PVA	Polyvinyl alcohol
RDP	Resorcinol di(phenyl phosphate)
REACH	Registration evaluation authorization and restriction of chemicals
SEM	Scanning electron microscope
SMC	Sheet moulding compounds
TDI	Toluene diisocyanate
TEP	Triethyl phosphate
TGA	Thermogravimetric analysis
THF	Tetrahydrofuran

THR	Total heat release
TPP	Triphenyl phosphate
TPS	Thermoplastic starch
TPU	Thermoplastic polyurethane
TTI	Time to ignition
UPCS	Urea salt of chitosan phosphate
XPS	X-ray photoelectron spectrometry
XRD	X-ray diffraction

Introduction

Flame retardancy is a property desired for many applications, in various fields as textile, wire and cable, transport, electric and electronic equipment, building and so on. Since polymers have invaded our everyday life, the fire hazard has grown. Indeed, most polymers, and especially the most used ones, burn vigorously and release large amounts of heat, promoting the propagation of fire. Fires cause numerous deaths and economic losses every year with significant differences from a country to another one. These differences are due in particular to fire regulations which are more or less severe. But, overall fire regulations are becoming increasingly stringent in developed countries and tend to converge. They led to use more and more two different strategies to prevent fire hazard. The first strategy is the use of active fire protective devices, as sprinklers or smoke detectors. The second one is to use materials contributing to fire as little as possible.

This strategy involves the use of flame retardants, which are additives incorporated into a polymer to reduce one or several aspects of its contribution to fire. Indeed, the fire hazard can be assessed from a couple of parameters as ignitability, heat release, flame propagation, smoke opacity and toxicity… Flame retardants are the first class of additives for polymers from turnover point of view. Flame retardants differ in their chemical nature, their mode-of-action and the amount to be incorporated. Halogen-based flame retardants act mainly in gaseous phase, by preventing the combustion. They can be very effective at low content but they produce toxic gases and a lot of smoke. They are gradually replaced by other flame retardants and research about them is very limited since at least 10 years. Therefore, their applicability to biobased materials has been little documented.

Mineral hydrated fillers (mainly aluminum trihydroxide (ATH) and magnesium hydroxide (MDH)) must be used at high content to be effective. Quite often the fraction of these flame retardants exceeds 50 wt%. Such high contents decrease mechanical properties and lead to processing issues. They release water through endothermic decomposition which cools the materials. Water dilutes fuel gases. Moreover mineral particles (alumina from ATH or magnesium oxide from MDH) accumulate on the surface as the polymer is burnt. This mineral layer can act as barrier protection slowing down the decomposition.

Some phosphorus-based flame retardants act as flame inhibitor but most often they behave as charring promoters. Charring decreases the amount of fuels sustaining the flame. Moreover, char accumulates at the surface and can create a barrier layer. When this layer is able to swell, it is particularly efficient to limit the heating of the underlying material. Such flame retardant systems are called intumescent and are often based on ammonium polyphosphate (APP). The content of phosphorus-based flame retardant systems is often ranged from 15 to 30 wt%. Probably, these flame retardants are the most studied ones by research community.

Nanoparticles are seldom used alone as flame retardant. But they can act as very effective synergist at very low content (1–5 wt%) if they are well dispersed (i.e., at a nanometric scale). They allow especially decreasing the peak of heat release rate measured in cone calorimeter test by improving the barrier effect. There is a great variety of nanoparticles but the most studied ones are clays (especially montmorillonites) and carbon nanotubes.

There are numerous fire tests to assess the contribution of a material to fire hazard. In fact, each regulation defines one specific test or a specific couple of tests for this assessment. Moreover, there are not general correlations between fire tests. Some materials can pass successfully one test and be unrated in another one. Four main tests are commonly used in research laboratories to assess the fire performance of a material. These four tests are amply cited in this book and then we describe them briefly in this introduction. Limiting Oxygen Index (LOI) is a simple test measuring the minimum oxygen content in atmosphere to ensure a self-sustained flame. The sample is positioned vertically and is ignited using a flame applied on its top surface. A material with a LOI lower than 21 (i.e., the oxygen content in air) burns easily when igniting, even without additional heat input. Higher is the LOI, better is the flame retardancy.

UL94 is a common and severe fire test. The sample is positioned vertically and a flame is applied on its bottom surface during 2 x 10 seconds. The duration of flaming period is recorded as well as the flame propagation and the dripping. V-0 rating corresponds to a material able to flame out almost as soon as the flame is removed without flaming drips. It is the best ranking. If the material needs more time to flame out, it can be V-1-rated. V-2-rating corresponds to materials requiring more time or promoting flaming drips. If the material does not flame out, it is non-rated.

Cone calorimeter is a useful test because it provides many data about the fire behavior for various fire scenarios. Typically, about 4 mm-thick sheet is submitted to a controlled heat flux in presence of a spark igniter (up to ignition). Most often the sheet is positioned horizontally. The heat flux is constant on its top surface and can be ranged from 10 to 100 KW/m^2 (most often heat flux of 35 or 50 kW/m^2 are chosen). The test is well ventilated (air flow 24 L/s). The oxygen concentration in the air flow is continuously recorded during the whole test. The combustion consumes oxygen and then the measurement of oxygen concentration allows assessing the intensity of burning. Indeed, empirical Huggett relation states that a consumption of 1 kg of oxygen corresponds to a heat release of 13.1 MJ whichever the materials. The peak of heat release rate (in W/m^2) is especially considered as the main parameter to assess the fire hazard. Mass loss, CO and CO_2 productions as well as smoke opacity are also measured.

Pyrolysis-combustion flow calorimeter (PCFC), also called microscale calorimeter, does not provide a complete overview on the fire behavior of a material. Nevertheless it can provide some useful data especially when only little amount of samples are available in the first steps of materials development. The sample (several milligrams) is pyrolyzed under nitrogen at a constant heating rate (1 K/s) up to 750 °C. The pyrolytic gases are sent to a combustor. In the presence of oxygen in excess and at high temperature (900 °C), the gases are fully oxidized and the heat release can be calculated according to Huggett relation. Main data provided using this method are intensity (in W per gram of sample) and temperature of the peak of heat release rate and the total heat release.

Reflections on sustainable development carried out from the 80s have brought up the industrial chemistry to question about its activities and to propose from the early 90s the concept of Green Chemistry that aims at reducing the hazardousness of chemicals. Green chemistry is based on 12 principles that have been defined by Anastas and Warner. One of these principles is the use of renewable raw matters. This way has raised a great interest in the field of polymers for the last 15 years. This growing fad is motivated by numerous reasons. In the first instance, alternatives to fossil resources (especially oil) have to be found since reserves are expected to progressively dry up and the control of oil deposits is a source of international confrontations. Second, environmental issues related to the manufacturing and use of polymeric materials have also to be solved. The expected effects are on one side the limitation of impacts associated to oil-based material production and on the other side a better management of the end of life of products, composting, and biodegradation being possible responses.

Therefore, a great trend in research and development field is to promote biobased materials. Biobased polymers, reinforcements (i.e., fibers) for composites, and additives are extensively studied. Researchers have also started to study the fire behavior of these new polymers or composites or provided by biobased flame retardants. This book reviews three aspects of these researches.

The first chapter is devoted to the flame retardancy of biobased polymers and mainly poly(lactic acid) (PLA), surely the main biobased polymer studied at present. In particular, after a short introduction about biobased polymers, this chapter discusses the applicability of usual flame retardant solutions (i.e., already developed for oil-based polymers) to PLA. Mainly nanoparticles and phosphorus-based flame retardants (especially ammonium polyphosphate) are reviewed. Also, innovative flame retardant systems involving new phosphorus-based molecules or macromolecules are detailed. A part of these compounds may act as reactive components.

When looking at all the commercially available polymers, it is obvious that many of them are not used in their pristine state. A great majority of these plastics contains additives that bring additional functionalities to materials: plasticizers, lubricants, antistatics, thermal stabilizers, UV stabilizers, anti-oxidants, dies, reinforcements…In order to have a consistent green chemistry approach, it is convenient to design additives for polymers (including flame retardants) that are based on renewable resources as well. The second chapter reviews how renewable resources have been used for the development of biobased flame retardants. Some of them correspond to mainly academic works, but other ones are available in large quantities and are cost-efficient.

Replacing glass or carbon fibers by natural fibers is still in line with the requirements of sustainable development. But ligno-cellulosic fibers are organic and contribute to heat release. Therefore the flame retardancy of biocomposites must be studied with a special emphasis. The third chapter discusses the flame retardancy of composites filled with natural plant-based fibers and compares the influence of ligno-cellulosic fibers on the flame retardancy to that of glass fibers. The strategy of flame retarding the reinforcement rather than the matrix is also assessed.

Of course, a fully biobased polymeric materials is a desired objective. Nevertheless, two remarks must be drawn. First, even if bioresources are initially used, they are often modified to obtain a suitable material. The fraction of biobased carbon in the final material can be lower than 1. In some cases, the final synthesized polymer or flame retardant is more oil-sourced than biobased. Second, "biobased" does not mean "green". Plant production consumes fertilizers, water and needs many energy-demanding steps. Therefore, the other principles of green chemistry may not be fulfilled. Only life cycle analysis may allow assessing the whole impact of oil- or biobased materials on the environment. Such life cycle analyses are rather scarce due particularly to non-available information. The two issues cited above are not included in this book because data are insufficient. More research is needed to address correctly these issues.

Chapter 1
Flame Retardant Biobased Polymers

1.1 Introduction

Environmental concerns arising from the limits to the waste management of plastics have entailed a strong development of biobased and biodegradable polymers for a wide range of applications. Tailoring new plastics and composites within a perspective of sustainable development aims to create an environmentally safe alternative to oil based polymer materials. Different categories of these polymers can be distinguished according to their complete or only partial renewable character as well as their ability to biodegrade. Shen et al. [1] presented a basic classification of the main fully or partially biobased polymers regarding their origin and biodegradability (Fig. 1.1).

A part of these polymers are agro-polymers (e.g. polysaccharides such as starch) obtained from biomass by fractionation. Another family corresponds to polyesters, obtained, respectively by fermentation from biomass or from genetically modified plants (e.g. polyhydroxyalkanoate: PHA) and by synthesis and fermentation from monomers obtained from biomass (e.g. polylactic acid: PLA). Other polyesters, are mainly synthesized by petrochemical processes (e.g. polycaprolactone: PCL, polyesteramide: PEA, aliphatic or aromatic copolyesters). Nevertheless, in some cases, some of them contain often biobased monomers such as succinic acid. Polybutylene succinate is now produced from biobased compounds [2]. As a matter of fact, there is a global tendency to replace significant oil-based polymers fractions by biobased ones to improve the environmental footprint of final products. In some cases, even for non-biodegradable polymers, polyols, diacids or alkyl chains comes from renewable products. For many industrial applications, fully or partially biobased polymers are already able to replace fully fossil-based polymers. This is mainly the case for short-life applications such as packaging. Nevertheless, more durable applications require specific issues to be addressed, such as resistance to ageing processes or flame retardancy, according to current regulations in building

© The Author(s) 2018
R. Sonnier et al., *Towards Bio-based Flame Retardant Polymers*,
Biobased Polymers, DOI 10.1007/978-3-319-67083-6_1

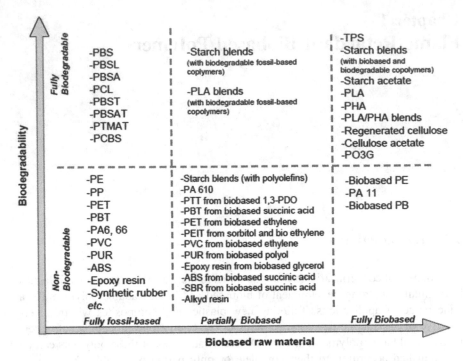

Fig. 1.1 Classification of biobased and biodegradable polymers, from Shen et al. [1]

and construction, transportation, electric and electric products as well as in textiles. Consequently, many research works have been carried out since around 2005 to transpose knowledge on flame retardants acquired from fossil-based polymers belonging to the same families of biobased polymers. In addition, original flame retardant systems devoted to these materials were also developed, taking into account their specific decomposition pathway. The most part of the research activity is focused on biopolyesters, and mainly on PLA. The main part of remaining research works is devoted to other biobased thermoplastics such as polybutylene succinate (PBS), polyurethanes or polyamide 11. In this chapter, the flame retardancy of biobased polymers similar to fossil-based ones (e.g. polyethylene) will not be considered, since it is abundantly reported in other reviews. The chapter is divided in three sections, the first one focuses on intumescent FR systems based on ammonium polyphosphate (APP), the second one to other phosphorous or nitrogeneous FRs systems, including new molecules. The last section is devoted to flame retardancy promoted by the incorporation of only natural or synthetic nanoparticles, surface modified in many cases, as well as different types of inorganic compounds, including metallic hydroxides. Due to the predominant role of flame retarded PLA, each section reports mainly experimental results about this polymer.

1.2 Intumescent Flame Retardant Systems Based on APP

Regarding the different strategies of flame retardancy, intumescence is among the most outstanding. The phenomenon of intumescence is based on the action of a polyacid which can dehydrate a charring agent, which can be the polymer matrix itself [3]. These components are completed by an expansion agent, in order to form an expanded foamed charred structure, able to limit mass and heat transfer when the material is heated. Different kinds of components can be used to fulfil each role in intumescent flame retardant systems (IFR), but in most cases, ammonium polyphosphate is used as polyacid source.

1.2.1 Use of Biobased Components with APP in PLA

Many researches were carried out from 2008 about the obtaining of intumescent FR systems for PLA. Reti et al. [4] evaluated the performances of the usual ammonium polyphosphate (APP)/pentaerythritol (PER) system, then, it was substituted by Kraft lignin and starch and an experimental design methodology was used to investigate the properties of the multicomponent compositions. Cone calorimeter analysis at 35 kW/m² performed on $100 \times 100 \times 3$ mm³ confirms the formation of an intumescent structures and final residues corresponding at least to 50% of the initial sample mass and also a decrease in the peak of heat release rate (pHRR) of more than 60% for 60% PLA + 30% APP + 10% PER (wt%) (Table 1.1). Moreover, a value of 32 for the Limiting Oxygen Index was obtained for a composite containing 60% PLA, 12% APP, and 28% starch.

More recently, in order to limit the reaction between ammonium polyphosphate and starch during processing, Wang et al. [5] added microencapsulated ammonium polyphosphate (MCAPP) to PLA/starch. The interest of using microencapsulation is related not only to the improvement of APP compatibility with PLA but also to delay the reaction between acid and carbonization agent during processing. The same authors had previously incorporated starch and MCAPP in biodegradable poly (vinyl alcohol) and found that starch conferred an excellent FR effect on PVA [6].

The flame-retardant properties of PLA/starch biocomposites were evaluated by limiting oxygen index, UL-94 test, and PCFC: Pyrolysis Combustion flow

Table 1.1 Cone calorimeter data from Reti et al. [4] for pure PLA, blends of PLA with Ammonium Polyphosphate and pentaerythritol, or lignin, or starch

Composition	$t_{ignition}$ (s)	pHRR (kW/m²)	Final residue (wt%)
PLA	72	325	0
PLA/APP/pentaerythritol	75	117	67
PLA/APP/lignin	87	173	50
PLA/APP/starch	93	193	50

Calorimetry). In this technique, some mg of materials are pyrolised according to a very steep temperature ramp (typically 1 K/s) and the gaseous degradation products are burnt in a combustion [7]. Similarly as in cone calorimeter, HRR is registered (expressed in W/g) and pHRR is determined as well as Total Heat Released (THR). The Heat Released Capacity (HRC) is defined as the ratio of HRR to the temperature ramp.

The results of PCFC obtained from Wu et al. [6] showed that the peak of heat release rate and total heat release of PLA/starch biocomposites decreased dramatically compared with those of pure PLA. IFR could reach UL-94 V-0 and a high LOI of 41.0. The thermal degradation and released gases of PLA/starch/MCAPP composition were investigated using thermogravimetric analysis coupled with infrared spectrometry. The TGA results indicated that the addition of IFR into PLA could strongly increase char yields and thermal stability of the char at high temperature compared with neat PLA. These results were consistent with the data of dynamic FTIR. More recently, Cayla et al. [8] have associated APP with Kraft lignin as char promoter in the intumescent system. Different formulations were prepared by melt extrusion and then hot-pressed into sheets. The spinnability of the various composites was assessed. A PLA multifilament with up to 10 wt% of intumescent formulation was processed, and the fire behavior of PLA fabrics with lignin and APP was studied by cone calorimeter (samples of $100 \times 100 \times 2$ mm, irradiance of 25 kW/m^2) (Table 1.2).

This last blend exhibits the best fire reaction, but even if the pHRR is slightly higher in comparison to PLA with APP composite, better results are found for the other fire parameters of the composite. In particular, the time of ignition is closer to this of pristine PLA. Moreover, THR and MAHRE (Maximal Average of Heat Rate Evolved) are respectively decreased of 56 and 43% in comparison with PLA. Moreover the lignin/APP combination allows the charring to be increased and presents the highest residue.

Table 1.2 Cone calorimeter data on fabrics (from Cayla et al. [8]) about pure PLA and compositions containing respectively 5 wt% of ammonium polyphosphate, 5 and 20 wt% of lignin, and finally 5 wt% of lignin combined with ammonium polyphosphate

Composition	$t_{ignition}$ (s)	pHRR (kW/m^2)	THR (MJ/m^2)	MAHRE (kW/m^2)	Final residue (wt%)
PLA	126	232	25	62	4
5 wt% APP	90	143	18	42	27
5 wt% lignin	143	228	17	57	13
20 wt% lignin	119	315	24	78	14
5 wt% lignin + 5 wt% APP	112	157	11	35	36

Different developments have been made also from IFR systems based on APP and lignin. Zhang et al. [9] have used lignin–silica hybrids (LSHs) prepared by sol–gel method. With the addition of APP/LSH to PLA system, the morphology of the cone calorimeter char residue has obviously changed and better flame retardancy is achieved, compared with PLA/APP and PLA/APP/lignin. The LOI value of the composites containing APP/LSH at different ratios (2:1, 3:1, and 4:1) is higher than that of the composites containing APP/lignin at the same ratios, and most ternary composites containing APP/LSH can reach V-0 rating in UL-94 test. A continuous and dense intumescent charring layer with more phosphorus analyzed using XPS is formed in PLA.

Zhang et al. [10] have also combined microencapsulated APP with lignin (ratio not indicated) and different organomodified montmorillonites (OMMT). The flame retardant and thermal properties of the composites were evaluated by limiting oxygen index (LOI), vertical burning test (UL-94), and cone calorimeter (samples of $100 \times 100 \times 3$ mm, irradiance of 35 kW/m^2) (Fig. 1.2).

From the results, it could be seen that the sample containing OMMT with methyl tallow bis(2-hydroxyethyl) ammonium modifier (OMMT2) possesses better flame retardancy, such as lower peak heat release rate (pHRR) and higher LOI value of 35.3. The results of TG-FTIR show that the presence of all OMMTs catalyzed the degradation of PLA, and that the sample containing the above mentioned OMMT releases less flammable gas products than the samples containing the other OMMTs. In addition, the char residue analysis clearly shows that its incorporation can improve the char quality with much more compact and continuous morphology.

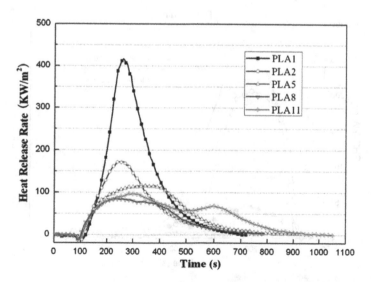

Fig. 1.2 Heat release rate as function of time, PLA1 (pure PLA), PLA2 (23 wt% IFR), PLA5 (21 wt% IFR + 2 wt% OMMT1), PLA8 (21 wt% IFR + 2 wt% OMMT2), PLA11 (21 wt% IFR + 2 wt% OMMT3), from Zhang et al. [10]

1.2.2 Use of Nanoparticles and Functional Fillers with APP in PLA

It has been shown in the end of the previous section that APP could be interestingly associated with biobased components and also with organomodified clays in PLA. It can be noticed that binary compositions of APP with nanoparticles were also investigated. On the whole, APP has been often used in association with charring, expansion agents and various types of nanoparticles for many polymers [11]. OMMT is the main category of nanoparticles involved in such systems. The combination of APP with melamine and OMMT has been successfully investigated by Fontaine et al. [12]. At first, it was shown that a PLA/APP/melamine (APP/Melamine 25:5 wt%) composition leads to a very low pHRR, around 40 kW/m^2. Tests were performed on $100 \times 100 \times 3$ mm^3 samples at an irradiance of 35 kW/m^2. Moreover, the addition of 1 wt% of OMMT to this IFR entails a nearby non-flammable material regarding the HRR values obtained at cone calorimeter (Fig. 1.3). This composition also allows a V-0 rating to be obtained at UL94 test (thickness of 3.2 mm). The use of 1 wt% carbon nanotubes instead of OMMT in the same IFRs seems less advantageous, even if a V-0 rating is similarly achieved.

Layered Double Hydroxides have been used as alternative layered nanoparticles to OMMT in research works proving the interest of lamellar nanoparticles regarding flame retardancy [13–15]. Wang et al. [16] have investigated PLA compositions with APP and pentaerythritol containing also 2 wt% zinc or magnesium aluminum layered double hydroxide (Zn-Al-LDH and Mg-Al-LDH). All composites have

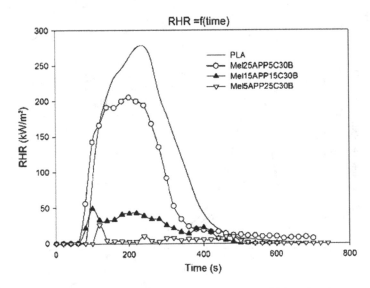

Fig. 1.3 Rate of heat released (HRR) as function of time for pure PLA and various compositions with a constant loading of 30 wt%: melamine + APP + organomodified montmorillonite (Cloisite 30B), from Fontaine et al. [12]

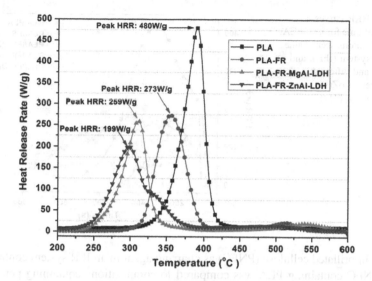

Fig. 1.4 HRR curves at PCFC as function of temperature for PLA, flame retarded PLA and flame retarded PLA containing 2 wt% of each LDH, from Wang et al. [16]

been prepared by melt-compounding directly at a global amount of 25 wt%. The morphology and burning behaviour of composites were investigated. HRR curves obtained at PCFC (1 K/s up to 700 °C in N_2) (Fig. 1.4) show a significant decrease of pHRR using particularly Zn-Al-LDH, nevertheless the temperature corresponding to the peak is dramatically reduced.

ZrP is another kind of layered nanoparticle which has been also previously investigated in various polymers such as epoxy resins, polyacrylamide and PET [17–20]. Liu et al. [21] have used an IFR system containing APP, an organo-modified α-zirconium phosphate (oZrP) and a charring agent [22]. oZrP was synthesized directly by a solvent thermal method using octadecyl dimethyl tertiary amine as organomodifier. TGA data have shown that oZrP could increase the residue. The addition of oZrP (1–3 wt%) to the flame retardant PLA at a global loading of 10 wt% increases the LOI and enhances the UL-94 rating with the best values obtained with a 2 wt% of oZrP. Incorporation of oZrP into PLA/IFR resulted in a significant reduction of HRR, THR and increased amount of char residues during combustion (Fig. 1.5) for cone calorimeter tests (samples of 100 × 100 × 6 mm^3 at an irradiance of 35 kW/m^2). LOI value could be increased from 30.5 to 35.5 with the substitution of 1 wt% of the IFR system for a global loading of 10 wt%, highlighting synergistic effects.

The flame-retardant mechanism of PLA/IFR/oZrP nanocomposites was ascribed both to the intrinsic flame-retardant mechanism of the IFR and to the catalyzed carbonization mechanism caused by oZrP.

POSS have been previously investigated as FR [23] or component of FR systems, for example with phosphinates [24] in PET. Fox et al. [25, 26] used POSS

Fig. 1.5 HRR curves as a function of time for neat PLA, PLA with intumescent flame retardant system (IFR), and with IFR and different percentages of oZrP, from Liu et al. [21]

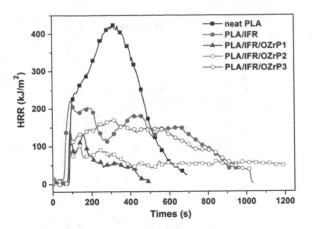

modified fibrillated cellulose (PNFC) as charring agent in an IFR system containing APP. PNFC containing PLA was compared to compositions containing pentaery-thritol or unmodified nanofibrillated cellulose (NFC). The use of PNFC resulted in the highest PLA molar mass of all flame retarded composites. In addition, PNFC formed a cross-linked network with APP when melt-blended with PLA, which reduced polymer degradation, decreased PLA crystallinity, reduced the melt vis-cosity, and improved composite stiffness relative to the neat extruded PLA. The addition of 15% by mass IFR reduced pHRR by 45% and THR by 20% (circular samples: 75 mm diameter, 3 mm thickness at 35 kW/m^2), regardless of the carbon source. PNFC led to reduced smoke production compared to NFC or pentaery-thritol. Finally, it was concluded that the PLA composites containing APP and PNFC exhibit the best tensile properties of all the intumescing composites studied. The cross-linked network formed between cellulose, POSS, and PLA helps to produce composites with superior flame retardant, rheological, and mechanical properties relative to other intumescent formulations.

Expanded graphite (EG) is often used as component of many FR systems, par-ticularly in polyurethanes [27]. Murariu et al. [28] have incorporated EG in PLA and studied the fire performance imparted. It has been shown that the addition of 6 wt% of EG leads to a reduction of about 30% of pHRR. A comparison of EG/PLA with pristine PLA reveals an important amount of char formed with EG, with a thickness directly correlated to EG percentage. Zhu et al. [29] have incorporated 15 wt% APP/EG(1:3) combinations. LOI values of 36.5 and V-0 rating in UL-94 tests were achieved, showing synergistic effects in comparison with APP or EG alone. Results from TGA and cone calorimeter (samples of $100 \times 100 \times 3$ mm^3 at an irradiance of 35 kW/m^2) (Table 1.3) proved that APP/EG combination could retard the degradation of polymeric materials above the temperature of 520 °C by promoting the formation of a compact char layer. This char layer plays the role of a thermal barrier, preventing further degradation, and resulting in reduced weight loss rate, final residue and better fire performance, taking into account pHRR reduction.

Table 1.3 Cone calorimeter data of PLA compositions, from Zhu et al. [29]: pure PLA and compositions containing respectively 15 wt% of ammonium polyphosphate and a combination of 3.75 wt% of expanded graphite and 11.25 wt% of ammonium polyphosphate

Composition	$t_{ignition}$ (s)	pHRR (kW/m^2)	THR (MJ/m^2)	Final residue (wt %)
PLA	60	272.0	65.6	4.0
15 wt% APP	70	208.4	46.1	31.0
3.75 wt% EG + 11.25 wt% APP	71	167.9	46.9	33.1

Various inorganic substances or fillers have been tested in combination with IFR systems, to achieve synergistic effects to improve the fire reaction of polymers, such as metal oxides [11, 30]. Lanthanide oxides were tested by Feng et al. [31].

1–4 wt% of La_2O_3 were associated to an IFR system made of APP and a charring agent containing triazine and benzene rings for a global loading of 20 wt%. Fire retardancy was assessed using PCFC and a fire index was defined according to the method proposed by Yang [32]. Morphology and composition of cone calorimeter residues as well as catalytic efficiency (CAT-EFF), defined as the increment in LOI per mass percent of metal ion were investigated. It is shown that IFR with 1% La_2O_3 presents the highest CAT-EFF, and the LOI value of the composite increases from 32.8 to a maximum value of 46.0 for 2 wt% of La_2O_3 (Table 1.4).

La_2O_3 enhanced the fire retardant performance of PLA/IFR. Hence, a reduction of heat release rate (HRR), total heat release (THR), and heat release capacity (HRC) of the PLA/IFR (Table 1.5) is noticed. The TGA test reveals that La_2O_3 enhances the thermal stability and changes the degradation behavior of the PLA with IFR, and increases the char residue. La_2O_3 helps to form more continuous and compact intumescent char layer on the surface. Higher graphitization degree of the residue char is observed. EDS and XPS techniques highlight that more P and O are present in the char layer to form more crosslinked structure to improve its barrier properties.

New developments for IFR systems in PLA stem also from the use of original charring structures combined with APP such as in the above mentioned work of

Table 1.4 Catalytic effect of La_2O_3 from LOI data, from Feng et al. [31]

La_2O_3 (wt%)	LOI (%)	Δ LOI (%)	La (wt%)	CAT-EFF
PLA	26.7	–	–	–
0	32.8	6.1	–	–
1	42.2	15.5	0.85	18.24
2	46.0	19.3	1.7	11.35
3	45.7	19	2.55	7.45
4	42.7	16	3.4	4.71

Table 1.5 PCFC data of PLA compositions: pure PLA, PLA with IFR, and PLA with IFR and La$_2$O$_3$, from Feng et al. [31]

Composition	HRC (J/K/g)	pHRR (W/g)	THR (kJ/g)	T$_{pHRR}$ (°C)	F$_{index}$
PLA	508	504.4	17.0	380	1.36
PLA/IFR	402	396.8	14.7	373	1.04
PLA/IFR/1 wt% La$_2$O$_3$	373	369.9	14.5	374	0.94

Feng et al. [31]. Ke et al. [33] synthesized a hyperbranched polyamine charring agent (HPCA), also derived from triazines. Different compositions of APP and HPCA were prepared, at a global loading of 30 wt%. The results show that the IFR system exhibit both excellent flame retardant and anti-dripping abilities for PLA, particularly for APP/HPCA = 1.5 wt/wt ratio. LOI value of 36.5 was attained for this composition. The synergistic effect between APP and HPCA was ascribed to a better ability to promote char formation.

1.2.3   Other Components in Intumescent Systems with APP for Biobased Polymers

Another strategy consists in using specific polymers as char promoters. Shabanian et al. [34] incorporated a novel polyamide (PA) containing aromatic and aliphatic groups, which was synthesized from a polycondensation reaction of adipic acid with 4,4-diaminodiphenyl sulphone. The flammability of PLA and PA were studied by PCFC (1 K/s). The thermal stability and flame retardant properties of PLA-PA composites were investigated by TGA and cone calorimeter. PA exhibited a synergistic effect on flame retardancy of PLA at the mass ratio of 2:1 for APP and PA at a global percentage of 15 wt%, showing delayed time to ignition (TTI), but only a slight decrease in pHRR at cone calorimeter test (samples of $100 \times 100 \times 3$ mm^3, irradiance of 25 kW/m^2). The use of a new composition, with the same APP/PA ratio, a global loading of 15 wt%, containing 1 wt% OMMT, allowed pHRR, HRR value to be significantly reduced and duration of the combustion to be strongly increased.

In the work of Bocz et al. [35], stacking of highly crystalline PLA fibres with amorphous PLA films was carried out. APP combined with OMMT (10:1 ratio) was incorporated between the matrix layers. As low as 16 wt% APP/OMMT loading proved to be sufficient for achieving self-extinguishing behaviour. UL94 V-0 rating was obtained. 50 and 40% reduction of pHRR and THR were achieved respectively. Despite PLA fibres did not seem to improve intrinsically fire performance and increase the char formation, their incorporation seemed to confer a better mechanical consolidation of the residual material during thermal degradation.

All previous articles reported are related to the incorporation of the IFR system in the bulk of PLA. Recently, fire protection of plastics and fabrics through surface coatings has been renewed by the use of layer by layer (LbL) assembly techniques in which thin films of various FR components are successively deposited on the surface [36]. Since PLA is a comparatively brittle and rigid polymer material with low elongation at break, applications in many end-use areas are hindered [37]. In addition, the deterioration on toughness of PLA and PLA composites is exacerbated with the use of many FR incorporated in the bulk as micronic particles. Jing et al. [38] developed a novel biobased hybrid with a core-shell structure made of a core of APP covered with an organic shell progressively built by LbL assembly of a novel biobased polyelectrolyte (BPE) and polyethylenimine (PEI). It has been shown that hybrid can simultaneously enhance the flame retardancy and toughness of poly-lactic acid (PLA) with very high efficiency. The flame retardant PLA composite can pass UL94 V-0 rating at the hybrid loading of 10 wt%. The effectiveness of flame retardant performance was ascribed by the authors to mechanisms in both gas and condensed phases. Moreover, the PLA flame retarded with 10 wt% hybrid exhibits ductile fracture with an elongation at break of 27.3%, which is much higher than that of pristine PLA (8%).

As mentioned in the introduction, the use of FR systems in other biobased polymers has not been significantly reported. For APP-based IFR, some works can be mentioned in PA11 and PBS. In the work of Levchik et al. [39], only the influence of APP on the fire behaviour of PA11 was investigated. A mechanistic and kinetic study was carried out and showed that APP strongly modifies the thermal degradation pathway of PA11 by lowering its decomposition temperatures and changing the composition of the released volatile products. α- and β-unsaturated nitriles are the main volatile decomposition products in the presence of APP. Minor amounts of hydrocarbons are evolved from neat PA11, but their formation is suppressed in presence of APP. Despite a charring process with intumescence occurring in parallel with the volatilization of PA11, APP seems not to have a significant influence on the yield of residue formed. The mechanism proposed for the interaction of APP with PA11 involves the formation of inter-mediate phosphate ester bonds which decomposes later. The fire-retardant action is mainly ascribed to the intumescent behaviour of the char.

More recently, our research group has worked on flame retardancy of PBS and its composites in which APP was associated with different natural fibres and par-ticularly flax [40] and also with natural nanoclays such as halloysite and sepiolite [41]. Ternary compositions containing APP, sepiolite and alkali lignin were also prepared. The formation of metallic phosphates during cone calorimeter testing (samples of $100 \times 100 \times 4$ mm^3, irradiance of 50 kW/m^2) allows a protective and expanded layer to be formed with significant reduction in pHRR (Table 1.6). Moreover, equimassic ternary composition of the three component of IFR system at a global loading of 20 wt% leads to the highest yield of residue and phosphorus retained in it, as well as excellent flame retardant performance.

Combinations of IFR systems with engineering nanoparticles in PBS seem promising as well. Wang et al. [42] incorporated 0.5, 1 and 2 wt% of graphene in an

Table 1.6 Cone calorimeter data of PBS compositions with APP (A), Lignin (L), Sepiolite (S), from Dumazert et al. [41]. LSA corresponds to equimassic percentage of 6.67 wt% for each component

Sample	Time of ignition (s)	pHRR (kW/m^2)	THR (MJ/m^2)	EHC (MJ/kg)	MAHRE (kW/m^2)	Xchar (%)	Total smoke released (m^2/m^2)
A20	42	367	94	20.3	398	20.4	1042
L10	31	402	98	19.1	292.6	12.4	1024
L5A15	34	513	97	19.3	292	12.8	987
L5S5A10	33	329	106	20.9	235	11.8	1064
LSA	32	267	104	20.7	174.1	13.0	603

IFR system containing APP and melamine (ratio 2:1). The limiting oxygen index (LOI) values increased from 23.0 for the pure PBS to 31.0 with 20 wt% IFR loading including 2 wt% graphene and the UL-94 V0 rating was obtained. The addition of graphene exhibited excellent antidripping properties due to the enhancement of melt viscosity.

1.3   FR Systems Based on Other Phosphorous and Nitrogenous FRs

Many phosphorous-based FR systems are commercially available according to the different chemical structures of phosphorous FRs and the diverse combinations with other FR components. Other inorganic phosphates than APP are of interest as well as organic phosphates, phosphonates, phosphinates, specific chemical structures such as DOPO (9,10-dihydro-9-oxa-10-phosphaphenantrene-10-oxide) and its derivatives, also phosphazenes...

In addition, novel molecules containing phosphorus have been synthesized and investigated as FRs in usual polymers. Another route consists in chemical modifications of polymers designed to graft or to introduce phosphorous chemical groups in the macromolecular chain to obtain what is called reactive FRs (vs. additive ones).

1.3.1   Systems Based on Other Types of Phosphates

Melamine phosphate (MP) or polyphosphate (MPP) are often used as alternative compounds to APP, for example in unsaturated polyesters, epoxy resins or polyamides. MP was associated with pentaerythritol phosphate (PEPA) and a polyhedral oligomeric silsesquioxanes (trisilanolisobutyl POSS) by Song et al. [43] (Fig. 1.6).

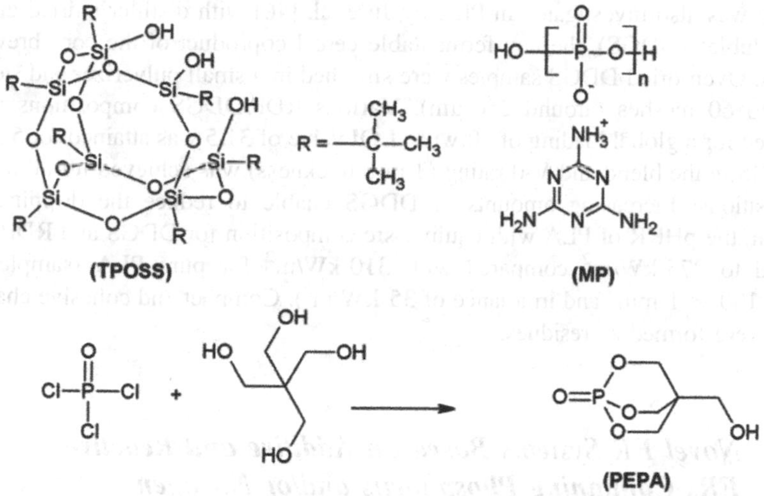

Fig. 1.6 Scheme of trisilanolisobutyl POSS, melamine phosphate and PEPA, from Song et al. [43]

POSS amounts of 1–5 wt% were used for a global loading of 25 wt% of IFR system. LOI value of 36% and V0 rating was achieved for 5 wt% POSS. The presence of MP and PEPA can promote the formation of char layer, while TPOSS is oxidized to form silica covering the char layer, which improved the thermo-oxidation resistance of the char.

Murariu et al. [44] associated melamine phosphate (MP) with $CaSO_4$. Hydrated calcium sulphate (β-anhydride II noted AII) as by-product of lactid acid processing was used. Co-addition of AII (25 wt%) and MP (15 wt%) allowed only a V-2 rating to UL 94 test, while a V-0 rating was attained by increasing the loading of MP at 25 wt% in PLA. Moreover, LOI of PLA with 25 wt% AII (LOI = 23%) increased above 33% with the co-addition of 25% MP into PLA–AII.

Among the organic phosphates, RDP (Resorcinol Diphenyl Phosphate) is currently used in various polymers such as PC, ABS or ABS/PC. Pack et al. [45] used coated starch with RDP, OMMT and halloysite in PLA and an aliphatic-aromatic co-polyester (Ecoflex, F BX 7011), provided from BASF. RDP is presented as a biodegradable compound with good thermal stability. This combination entails a self-extinguishing behaviour ascribed to the formation of shell-like chars, able to prevent molten polymer against dripping. The mechanical properties of the charred structures were studied using nano-indentation. It is considered that a relevant elastic behaviour for the char formed from PLA/Ecoflex blend containing the coated starch with nanoparticles and RDP could keep the internal pressures built up with decomposed gases from molten polymers. When exposed to the heat source, the clays prevented phase segregation, while the RDP starch migrated to the surface where it reacted and formed a hard shell that prevented dripping.

RDP was also investigated in PLA by Ju et al. [46] with distiller's dried grains with solubles (DDGS), the non-fermentable cereal coproduct of the corn brewing process. Oven-dried DDGS samples were smashed in a small pulverizer and sieved down to 60 meshes (around 250 μm). Various RDP/DDGS compositions were produced for a global loading of 30 wt%. LOI value of 31.5 was attained for 5 wt% of DDGS in the blend and V-0 rating (3 mm thickness) was achieved for all mixed compositions. Increasing amounts of DDGS enable to reduce the dripping. In addition, the pHRR of PLA with equimassic composition for DDGS and RDP was reduced to 275 kW/m^2 compared with 310 kW/m^2 for pure PLA (samples of $100 \times 100 \times 1$ mm^3 and irradiance of 35 kW/m^2). Compact and cohesive charred layers were formed as residues.

1.3.2   Novel FR Systems Based on Additive and Reactive FRs Containing Phosphorus and/or Nitrogen

Jing et al. [47] used diphenolic acid, a plant derived compound, to synthesize a biphosphate containing two types of phosphate groups (Fig. 1.7).

UL-94 V-0 rating and LOI value of 27.4 for PLA were achieved for a FR loading content of 16 wt% (Table 1.7). Besides, for cone calorimeter tests performed at 35 kW/m^2, pHRR was reduced from 418 to 336 kW/m^2 and THR from 60 to 50 MJ/m^2.

Chen et al. [48] synthesized a hyperbranched polyphosphate ester (HPE) and investigated the flammability and thermal stability of flame retardant PLA using limiting oxygen index (LOI), UL-94 vertical burning test and PCFC. 20 wt% HPE

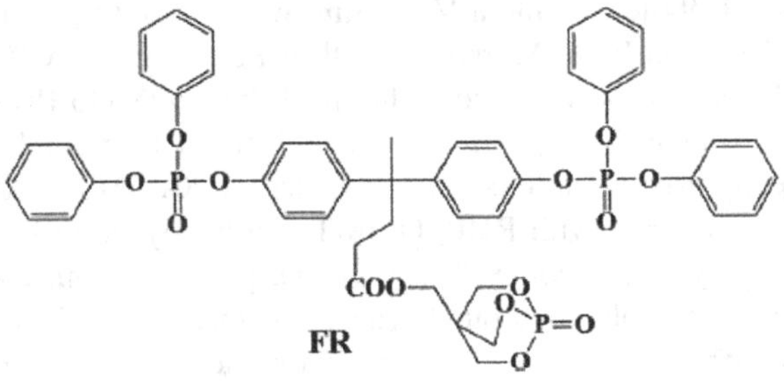

Fig. 1.7 Scheme of biphosphate FR, from Jing et al. [47]

Table 1.7 LOI and UL94 rating of PLA with biphosphate, from Jing et al. [47]

Composition	LOI (%)	UL 94 rating	Dripping	Cotton ignition
PLA	20.0	Failed	Drip	Yes
8 wt% biphosphate	24.2	V2	Drip	Yes
12 wt% biphosphate	25.4	V2	Heavy drip	Yes
16 wt% biphosphate	27.4	V0	Heavy drip	No

in PLA leads to a LOI value of 35%. Flame retardant PLA containing the highest percentage of HPE exhibited relatively lower thermal stability at lower temperature while higher thermal stability at elevated temperature and more char was formed compared with those containing a lower percentage of HPE. TGA coupled with FTIR showed that HPE catalyzed the degradation of PLA, and reduced the combustible gas products released during the thermal degradation process.

Phosphinates and particularly diethyl aluminium phosphinate (AlPi) have proved their efficiency in different polymers such as mainly PMMA, polyamides and PBT [49–51]. Their incorporation in PLA has led to interesting results but on the whole, the fire performance remains slightly lower than similar compositions using APP. Bourbigot and Fontaine [52] compared the fire performance of APP, MP, and two AlPi compositions with synergists in PLA at the same loading (up to 30 wt%). Highest LOI value (around 34) is obtained for the composition with only APP for a loading higher than 20 wt%. More recently, Isitman et al. [53] investigated the role of the geometry of different nanoparticles (nanosilica, halloysite and OMMT) on the flame retardancy of PLA containing AlPi. Incorporation of AlPi increases the fire retardancy of PLA by improving LOI and UL94 performances and reduces pHRR and peak of mass loss rate (PMLR). It appear that silica (0-D: pseudo-spherical) and halloysite (1-D: tubular) nanofillers does not entail outstanding improvements in fire retardancy regarding fire parameters, OMMT (2-D: lamellar) allows a large reduction in fire risks to be achieved. The underlying mechanism enabling enhanced fire retardancy by the 2-D nanofiller is explained by the authors by the migration and accumulation of aluminosilicate platelets on the exposed surface. Rapid migration of exfoliated platelets facilitates the formation of a thick and rigid residue composed of aluminium phosphate microspheres reinforced at the nanoscale by montmorillonite platelets, establishing an effective barrier to heat and mass transfer during combustion.

Synthesis of novel phosphinates is an innovative route to improve fire retardancy. Poly (1, 2-propanediol 2-carboxyethyl phenyl phosphinate) (PCPP) was prepared by Lin et al. [54] (Fig. 1.8).

Various poly(lactic acid) (PLA) blends were prepared by direct melt compounding with PCPP as flame retardant and plasticizer. LOI value increased from 19.7 for neat PLA to 28.2 for the PLA with 10 wt% PCPP. The flame retardant and the mechanical properties are improved simultaneously for a loading of PCPP in the composites up to 15 wt% (Table 1.8). The morphology of the residual material and the P content measurement indicate that condensed-phase and gas phase flame-retardant mechanism are involved.

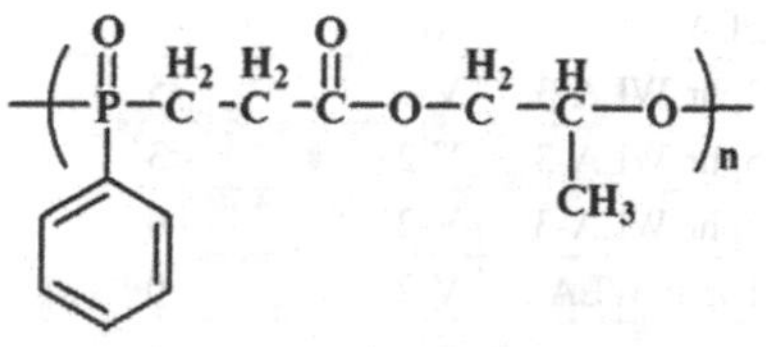

Fig. 1.8 Scheme of PCPP, from Lin et al. [54]

Table 1.8 UL-94 rating and LOI values for PLA containing various percentages of PCPP, from Lin et al. [54]

Composition	UL-94 rating	First ignition (s)	Cotton ignition	Observed dripping	LOI (%)
PLA	NR	>10	Yes	Drip	19.7
3 wt% PCPP	V-2	<10	Yes	Drip	26.6
5 wt% PCPP	V-0	<10	No	Drip	28.0
7 wt% PCPP	V-0	<10	No	Heavy dripping	28.0
10 wt% PCPP	V-0	<10	No	Heavy dripping	28.2

Phosphonates have been frequently used to flame retard textiles and commercial cyclic phosphonates are available such as AFLAMMIT PE (Thor) [55].

This compound was applied for the finishing of knitted fabric of PLA. It was found that the finished fabric exhibited durable flame retardancy regarding after-flame time and char length tests. Very recently, Cheng et al. [56] applied a cyclic phosphonate ester flame retardant to improve the flame retardancy of PLA nonwoven fabric by a pad-dry-cure technique. The effects of curing temperature and flame retardant percentage on the flammability of PLA fabric were studied. LOI of 35% was achieved for treated PLA fabric whereas 26.3% only for the untreated fabric. No significant difference was found between the different compositions in PCFC tests. TGA indicated the formation of only a very small amount of char during the thermal degradation of the treated fabric. SEM-EDS analysis showed a decrease in the P content of the fabric after burning, indicating a predominance of a gas-phase flame-retardant mechanism during combustion. An aryl polyphenyl phosphonate (noted WLA-3) in PLA, incorporated through direct melt compounding, was investigated by Wei et al. [57]. V-0 ratings for UL-94 test containing 7 and 10 phr of WLA-3 were achieved (Table 1.9).

However, cone calorimeter tests (samples of $100 \times 100 \times 6$ mm^3, irradiance of 35 kW/m^2) only showed a little decrease in HRR, pHRR and THR compared to neat PLA. TGA results showed that the PLA containing different amounts of WLA-3 presented more complex thermal decomposition pathway than pristine PLA.

Table 1.9 UL-94 rating and LOI values for PLA containing various percentages of aryl polyphenyl phosphonate (noted WLA-3), from Wei et al. [57]

Composition	UL 94 rating	First ignition (s)	Cotton ignition	Observed dripping	LOI (%)
PLA	NR	–	Yes	Drip	19.0
3phr WLA-3	V-2	<5	Yes	Drip	23.0
5phr WLA-3	V-2	<3	Yes	Heavy dripping	25.5
7phr WLA-3	V-2	0	No	Heavy dripping	25.0
10phr WLA-3	V-2	0	No	Heavy dripping	26.0

A specific strategy consists in integrating phosphorous groups in the backbone of PLA macromolecule. An inherently flame-retardant PLA noted PPLA was synthesized by Wang et al. [58] by direct polycondensation of L-lactic acid with ethyl phosphorodichloridate as chain extender. Only 5 wt% of PPLA added into PLA allow good flame retardant properties to be achieved. For a content of PPLA increased to 10 wt%, LOI value of 35% and UL-94 V-0 rating were achieved. Tests using cone calorimeter (samples of $100 \times 100 \times 6$ mm^3 and irradiance of 35 kW/m^2) show that lower pHRR value (436 vs. 337 kW/m^2) was observed in comparison with neat PLA.

Yuan et al. [59] incorporated a DOPO derivative in the PLA backbone via the chain extension of the dihydroxyl-terminated prepolymer with 1,6-hexamethylene diisocyanate (HDI) to obtain an IFR/PLA polymer which was then blended with the commercial PLA to prepare flame retardant PLA blends. Percentages of phosphorus in the chain were established at 1 or 2 wt% for various NCO/OH contents. TGA revealed that the char yield of IFR/PLA and flame retarded PLA blend above 400 °C was strongly enhanced compared to that of pure PLA. The LOI value was significantly improved to 29% when 1 wt% of phosphorus content was introduced and all IFR/PLA samples achieved V-0 rating in the UL-94 tests. Flame retarded PLA blends had an LOI value of 25-26 and UL-94 V-2 rating at 20 wt% of IFR/PLA content.

Different novel phosphorous compounds containing also nitrogen were synthesized in order to impart self-intumescent behaviour. Zhan et al. [60] synthesized a spirocyclic pentaerythritol bisphosphorate disphosphoryl melamine (SPDPM) (Fig. 1.9).

Different polylactide (PLA)-based flame retardant compositions containing SPDPM were prepared by the melt blending method (5–25 wt%). SPDPM is expected to integrate acid, char and gas sources to improve the flame retardancy of PLA. 25 wt% FR allows UL-94 V-0 and LOI value of 38% to be achieved while anti-dripping performance is increased. PCFC showed that SPDPM could significantly decrease the heat release capacity (HRC) and pHRR of PLA composites (Table 1.10).

The flame retardant mechanism of SPDPM in PLA was not only due to the intumescent protective charred layer resulting from combining acid, carbon and gas sources in SPDPM but also to the change of the degradation process of PLA according to analysis using TGA coupled with FTIR. It was shown that the low thermal stability of melamine segment results in the decrease of the decomposition

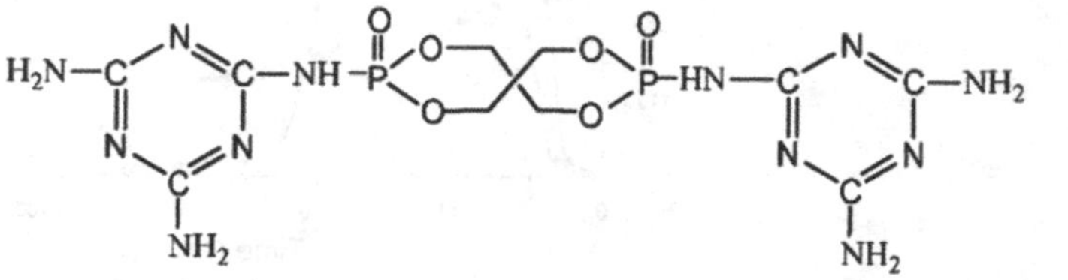

Fig. 1.9 Scheme of SPDPM, from Zhan et al. [60]

Table 1.10 PCFC data of PLA containing SPDPM, from Zhan et al. [60]

Compositions	HRC (J/g/K)	pHRR (W/g)	THR (kJ/g)	T_{pHRR} (°C)
PLA	492	475	16.4	390
5 wt% SPDPM	400	398	14.4	389
15 wt% SPDPM	291	292	12.2	377
25 wt% SPDPM	291	291	10.9	369

temperature onset. Another compound aiming to a similar behaviour was synthesized by Zhao et al. [61]: an intumescent flame retardant tris(2-hydroxyethyl) isocyanurate polyphosphate melamine (noted TPM). It was found that PLA with 25 wt% APP/TPM (0.5:1) led to a LOI of 36.5% and a UL-94 V-0 rating was achieved with no melt dripping. Cone calorimeter tests (samples of $100 \times 100 \times 3$ mm^3, irradiance of 50 kW/m^2) (Fig. 1.10) showed that pHRR and THR of PLA/TPM composites were dramatically reduced compared with that of pure PLA (255 vs. 655 kW/m^2). Further decrease was achieved through a substitution of a third of TPM by APP.

From the HRR curve the authors suggested that TPM could accelerate the char forming process. Moreover, TGA demonstrated that TPM effectively improved thermal stability of PLA. The development of a novel flame retardant polymer containing phosphorus, nitrogen, phenyl rings associated with ester groups on the backbone was synthesized by Liao et al. [62]. This polymer (noted PNFR) was blended with PLA at a percentage of 3–15 wt%. LOI value increased to 31.5% and UL-94 V-0 was attained for flame retarded PLA blend at a 3 wt% loading of PNFR. Even if dripping occurred, no ignition of the cotton wool was noted. SEM, Raman spectroscopy, XRD, XPS and FTIR results revealed that the compact and coherent char layer protecting the underlying material was partly graphitized. Phosphamide esters and phosphazene were also assessed as possible flame retardants for PLA. Li

Fig. 1.10 HRR curves of PLA and PLA with APP and TPM as a function of time, from Zhao et al. [61]

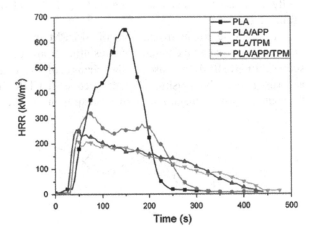

et al. [63] synthesized a hyperbranched poly(phosphamide ester) oligomer (noted HBPE) which was blended at a percentages of 2, 4, 6 and 10 wt%. It was shown that PLA blend with only 2 wt% HBPE exhibited excellent flame retardancy (LOI of 33% and V-0 rating at UL-94 test). Moreover, 10 wt% HBPE imparted PLA composites largely increased LOI value of 43%. The incorporation of 10 wt% HBPE increased CO production rate and increased time to ignition (TTI) in cone calorimeter. A novel phosphazene cyclomatrix network polymer poly (cyclotriphosphazene-co- pentaerythritol) (PCPP) was synthesized and blended with PLA (5–20 wt%) by Tao et al. [64] (Fig. 1.11).

UL-94 V-0 could be passed from 5 wt% PCPP. Absence of dripping was noticed from 10 wt%. LOI values increased as function of PCPP loading and a value of 28.2% is reached for 20 wt%. FTIR revealed that residual chars contain polyphosphoric or phosphoric acid, which indicated the probable carbonization reaction.

A last kind of compound also yet used as effective flame retardant in engineering plastics, such as other polyesters than PLA and also polyamides [15, 17] is aluminum hypophosphite (AHP). Tang et al. [65] incorporated 10–30 wt% AHP in PLA. LOI values increased continuously with the AHP content from 25.5 (10 wt%) to 29.5 (30 wt%). From 20 wt%, V-0 rating was attained (thickness of 3.2 mm). PCFC experiments showed that HRR, pHRR and THR of flame retarded PLA blends decreased significantly with the increase of AHP content (Table 1.11).

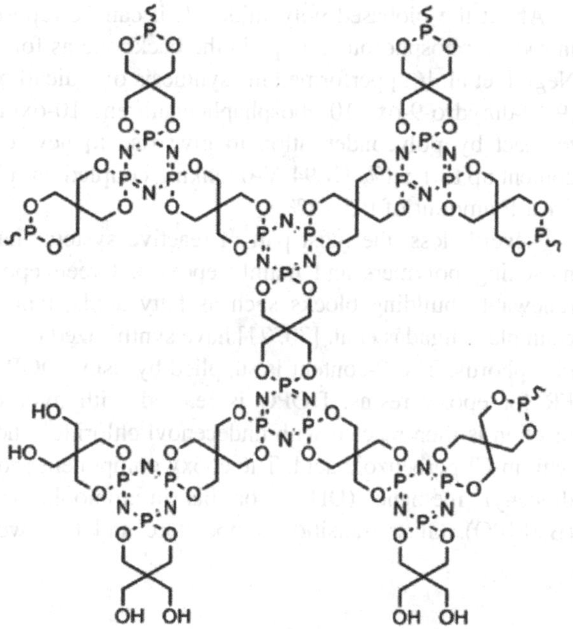

Fig. 1.11 Scheme of PCPP, from Tao et al. [64]

PCPP

Table 1.11 UL-94 rating and LOI values for PLA containing various percentages of AHP, from Tang et al. [65]

Composition	UL 94 rating	Dripping	t1/t2 (time to autoextinguishment)	LOI (%)
PLA	NR	Yes	–	19.5
10 wt% AHP	V-2	Yes	10.7/1.7	25.5
15 wt% AHP	V-2	Yes	8.3/3.4	27.5
20 wt% AHP	V-0	No	0.9/5.8	28.5
30 wt% AHP	V-0	No	1.1/2.2	29.5

The cone calorimeter test also confirmed that the addition of AHP results in a significant decrease in pHRR value of blends compared with pure PLA (550–250 kW/m^2). In a following work, the same authors [66] synthesized rare earth hypophosphites: La and Ce hypophosphites. Similar results in comparison with AHP were noticed, nevertheless, from cone calorimeter results this latter seemed more interesting.

The use of other phosphorous FRs than APP in other biobased thermoplastic polymers is insignificant. The above-mentioned FR synergism between phosphinates and metallic oxides was extended by Gallo et al. [67] to PHBV/PBAT [poly(3-hydroxybutyrate-co-3-hydroxyvalerate)/poly(butylene adipate-co-terephtalate) biopolyester blends. Also Bocz et al. [68] studied the interest of the use of phosphorous polyols as multifunctional plasticizers with a flame retardant effect to obtain flame retarded blends based on thermoplastic starch.

About the biobased polyamide 11, it can be reported that the same strategy to introduce phosphorous groups in the backbone as for biopolyesters can be applied. Negrell et al. [69] performed the synthesis of a dicarboxylic acid from DOPO cycle (9,10-dihydro-9-oxa-10-phosphaphenanthrene-10-oxide). This derivative was able to react by polycondensation to give rise to new copolyamides (coPA) with P content up to 1 wt%. UL94 V-0 ranking coupled with LOI value 40% was achieved for a P amount of 0.5 wt%.

Nevertheless, the most part of reactive systems have been developed for thermosetting polymers and mainly epoxies. Green epoxies can be produced using renewable building blocks such as fatty acids, tannins, catechol or cardanol for example. Lligadas et al. [70, 71] have synthesized epoxidized fatty acids containing phosphorus. The P-content is supplied by using DOPO, which is currently used as FR for epoxy resins. DOPO is reacted with hydroquinone and the product of reaction is then reacted with undecenoyl chloride. The next step is the epoxidation with m-Cl perbenzoic acid. The epoxy component produced is cured with diamino diphenyl methane (DDM) or bis-(m-aminophenyl) methyl phosphine oxide (BAMPO). Glass transition temperature and LOI were determined for the cured

Table 1.12 Glass temperature transition and LOI values of DOPO modified epoxy resins, from Lligadas et al. [71]

	Tg from Tan δ (°C)	LOI (%)
DDM-cured resin	108	31
BAMPO-cured resin	95	32

samples. From Table 1.12, it appears that DDM-cured sample is more advantageous to maintain the Tg of the resin.

The same research group synthesized acrylate and phosphine oxide- containing vegetable oils and prepared crosslinked materials [72]. The flame retardancy of the pristine and modified resins was assessed through LOI measurements. Unmodified resin exhibited a LOI value of 19.6 whereas the incorporation of 1.4 wt% led to a LOI of 21.2 and with 2.8 wt% a LOI of 22.4 was achieved. The slight improvement of FR properties was ascribed to the phosphorus content.

Similarly, reactive fire retardant PUs were developed from renewable building blocks. Zhang et al. [73] have prepared castor oil phosphate flame-retarded polyol (COFPL) derived by the reaction of epoxidized castor oil with diethyl phosphate and catalyst via a three-step synthesis. It was shown that for a content of P element of about 3%, the fire retardant incorporated in the castor oil chain increased thermal stability and LOI value of polyurethane foam can reach to 24.3 without any other flame retardant. Expanded graphite (EG), in addition to castor oil phosphate, provided excellent flame retardancy. The flame retardancy of PU foams determined with two different FR systems COFPL/EG and EG/COFPL/TEP (TEP: triethyl phosphate used as plasticizer) proved that EG/COFPL or EG/COFPL/TEP systems present a synergistic effect in castor oil-based PUFs. EG/COFPL PU foam exhibits a large reduction of peak of heat release rate (pHRR) compared to EG/GCO one at cone calorimeter tests (samples of $100 \times 100 \times 20$ mm^3 and irradiance of 35 kW/m^2) (Table 1.13). The SEM results show that the incorporation of COFPL and EG allows the formation of a cohesive and dense protective char layer, which limits the transfer of heat and combustible gas and thus increases the thermal stability of PUF. The enhancement in fire performance will expand the application range of COFPL-based polyurethane foam materials.

Table 1.13 Cone calorimeter data of castor oil based PU foams, from Zhang et al. [73]

	100phr COFPL + 30phr EG	100phr GCO + 30phr EG
pHRR (kW/m^2)	127.8	164.7
THR (MJ/m^2)	21.8	27.5
THR/mass loss (MJ/m^2/g)	1.82	2.29

1.4 Flame Retardant Systems Based on Functionalized or Modified Nanoparticles and Fillers

A wide range of nanoparticles has been used in various usual polymers. Various researchers have tried to transpose directly the use of such components in biobased polymers. Nevertheless, in some cases, specific surface modifications or functionalizations have been carried out to take into account the specific character of these polymers and particularly those of biopolyesters.

Liu et al. [74] have synthesized a novel functionalized α-zirconium phosphate (F-ZrP) by co-precipitation of a phosphate salt containing nitrogen and zirconyl chloride solution (Fig. 1.12).

The results showed that the addition of flame retardant F-ZrP slightly affects PLA's thermal stability, but significantly improved the flame retardancy of PLA composites. Table 1.14 shows that the functionalization leads to better LOI values, UL94 rating (3 mm thickness) and cone calorimeter data (samples of $100 \times 100 \times 3 \text{ mm}^3$ and irradiance of 35 kW/m^2) than with α-ZrP yet mentioned in the first section of this chapter.

Polylactide (PLA)/layered double hydroxide (LDH) films with good flame retardant property and transparency were synthesized by Ding et al. [75] using solution exfoliation and film casting method. The organic–inorganic interfacial interaction between PLA and NiAl-LDH was carefully modified by 2-carboxylethyl-phenyl-phosphinic acid (CEPPA) to promote a good dispersion of NiAl-LDH and to impart flame retardancy of PLA composites. Exfoliated NiAl-LDH/CEPPA (LC) structures homogenously dispersed in PLA matrices were achieved. All PLA/LDH films had good transparency even when the LC content was up to 10 wt%. The flame retardant effect is enhanced when LC percentage increases in PLA. THR measured using PCFC decreases continuously as function

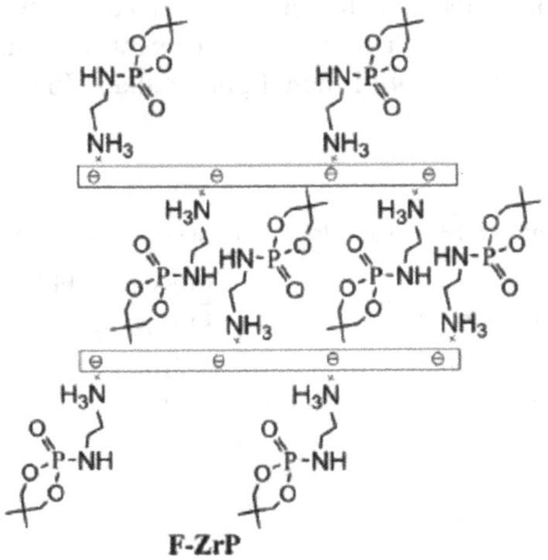

Fig. 1.12 Structure of F-ZrP, from Liu et al. [74]

F-ZrP

Table 1.14 LOI values, cone calorimeter data and UL94 rating of PLA with various percentages of F-ZrP in comparison with α-ZrP, from Liu et al. [74]

Composition	LOI (%)	UL 94 rating	pHRR (kW/m²)	Average HRR (kW/m²)	THR (MJ/m²)	Residual mass (%)
PLA	19	NR	424	239	146	0
5 wt% F-ZrP	24.5	V-2	268	178	112	5.6
10 wt% F-ZrP	26.5	V-0	246	120	85	14.3
20 wt% F-ZrP	29	V-0	203	115	77	25.6
5 wt% α-ZrP	21	NR	260	173	116	8.7
10 wt% α-ZrP	22.5	V-2	236	166	105	10.9
20 wt% α-ZrP	24.0	V-2	224	152	78	21.5

of the LC percentage. A value of 9.7 kJ/g is noticed for 10 wt% LC, whereas 12.0 kJ/g are measured for neat PLA.

Hu et al. [76] have prepared PLA nanocomposites by melt-blending with multiwalled carbon nanotubes (MWNT) and functionalized MWNT (MIP) with tri (1-hydroxyethyl-3-methylimidazolium chloride) phosphate (IP). Significant improvement in flame retardancy was observed for the PLA/MIP composite using cone calorimeter (samples of $100 \times 100 \times 4$ mm³ and irradiance of 35 kW/m²) since for 5 wt% MIP, the pHRR was reduced from 324 to 155 kW/m² and the THR decreased from 49 to 34 MJ/m² (Fig. 1.13). For this composition, LOI value of 26 was attained. In addition, TGA showed that the char residue was increased compared to PLA/MWNT. It was concluded that the catalytic charring effect of IP, the physical crosslinking effect promoted by MWNT, and the combined effect of both IP and MWNT (forming continuous and compact char layers) account for the fire performance of PLA/MIP composite.

Fig. 1.13 HRR curves of pure PLA, PLA containing pristine multiwall carbon nanotubes (MWNT), IP modifier and modified carbon nanotubes with IP (MIP), from Hu et al. [76]

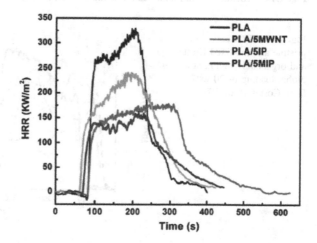

Biobased and biodegradable nanoparticles offer the opportunity to improve the ecological footprint of flame retardant biopolymers. To confer FR activity to these nanoparticles or nanofibres, chemical modifications can be performed, particularly by grafting or incorporating phosphorous groups.

Lignin, which was reported previously to have been combined with APP, was investigated as unique FR in PLA by Costes et al., regarding its nature [77]. Two different lignins, i.e. kraft and organosolv lignins were used. Influence of plant origin, extraction mode of lignins as well as their chemical modifications with phosphorus/nitrogen compounds (Fig. 1.14) were investigated. Fire properties and thermal behavior of PLA composites containing 20 wt% of both untreated and treated lignins (Lignin-PONH₄) were characterized by using cone calorimeter (Samples of $100 \times 100 \times 4$ mm^3 and irradiance of 35 kW/m^2), UL-94 (thickness of 3 mm) and thermogravimetric analysis (TGA).

Results showed that flame retardant action of untreated lignins corresponds to the formation of a char but also to a significant loss of thermal stability of PLA and to an important decrease of its time to ignition. Conversely, treated lignins limit PLA thermal degradation during melt processing as well as during TG experiments. Nevertheless, the treatment led to different HRR profiles at cone calorimeter than untreated ones (Fig. 1.15). No further decrease for pHRR was noticed, owing to the treatment.

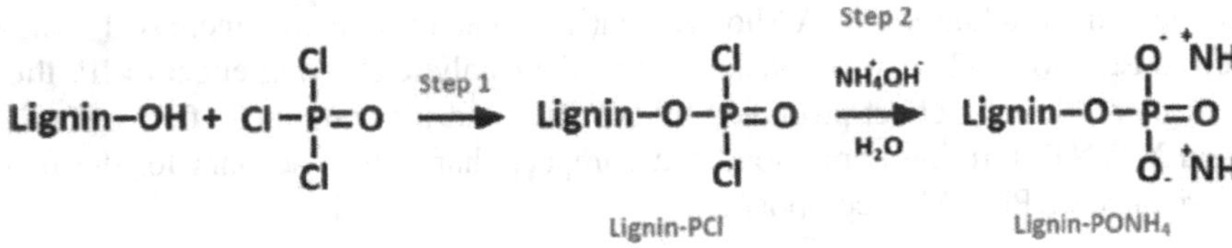

Fig. 1.14 Scheme of lignin modification using POCl₃, from Costes et al. [77]

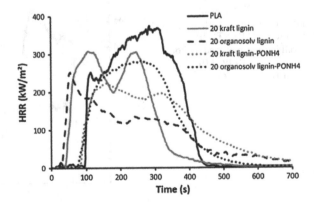

Fig. 1.15 HRR curves of pristine and modified Kraft and organosolv lignin, at a global loading of 20 wt%, from Costes et al. [77]

The effect of combining microcrystalline cellulose (MCC) or nanocrystalline cellulose (NCC) with phosphorus on thermal stability and flame retardant properties of PLA was also investigated by Costes et al. [78]. Phosphorus was incorporated either by chemical grafting on cellulose or by co-additive melt blending by using a biobased phosphorous agent such as aluminum phytate, also studied as flame retardant in a previous work [79]. In both cases, the charring effect of cellulose was enhanced. The use of 20 wt% of phosphorylated MCC (MCC-P) allowed reaching V-0 at UL-94 test but significant reduction of pHRR was only achieved when aluminum phytate was combined with MCC-P. The use of a phosphorylated-NCC was not required since an equimassic combination of aluminum phytate and NCC allowed a significant decrease of the pHRR to the value obtained when MCC-P was used in combination with aluminum phytate (around 240 kJ/m^2).

The association of two different types of nanoparticles can lead to complementary FR effects or even in some cases to synergistic ones if improvements of properties can be noticed for mixed compositions at a constant global loading. González et al. [80] incorporated OMMT and sepiolite at percentages of 5wt% and both nanoparticles together at 10 wt%. Dispersion of nanoparticles was investigated and intercalation of OMMT was noticed. It was shown that at two irradiances (25 and 35 kW/m^2) for cone calorimeter testing (samples of $100 \times 100 \times 3$ mm^3), each kind of nanoparticles allows the pHRR to be reduced (Table 1.15). Further decrease was achieved through the combination of OMMT and sepiolite, despite a loss of thermal stability observed using TGA.

Combinations of sepiolite (10 wt%) and multiwalled carbon nanotubes (2 wt%) in PLA were investigated by Dhanushka Hapuarachchi and Peijs [81]. Thermal gravimetric analysis (TGA) showed a significant increase of char yield. A reduction in heat release capacity (HRC) of 58% at PCFC was noticed for the ternary system based on sepiolite and MWNTs. Moreover, the nanocomposite showed a 45% reduction in pHRR when tested in the cone calorimeter (value of 265 kW/m^2).

Since nanoparticles have often be used as synergistic agents of metallic hydroxides in various polymers such as polyolefins [11, 82], OMMT/ATH combinations in PLA were proposed by Cheng et al. [83, 84]. On the whole, the incorporation of metallic hydroxides used as FR in polyesters can lead to hydrolysis phenomena during the first steps of thermal degradation. Moreover, hydrolysis can

Table 1.15 LOI and cone calorimeter data at an irradiance of 35 kW/m^2 for pure PLA, PLA containing 5 wt% of sepiolite, OMMT and equimassic blend of nanoparticles at a global loading of 10 wt%, from González et al. [80]

Composition	$t_{ignition}$ (s)	pHRR (kW/m^2)	tpHRR (s)	THR (MJ/m^2)	LOI (%)
PLA	394	74	74	22.8	20
5 wt% Sepiolite	419	79	79	29.3	20.6
5 wt% OMMT	467	82	82	24.9	20.8
5 wt% Sepiolite + 5 wt% OMMT	398	69	69	23.8	20.6

Fig. 1.16 HRR curves of
PLA, PLA with 40 wt%
ATH (B40), 38 wt% ATH
and 2 wt% of OMMT(B38),
35 wt% ATH and 5 wt%
OMMT(B35), from Cheng
et al. [83]

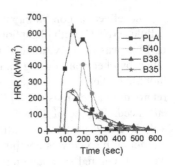

even occur during polymer processing. However, the thermal oxidative degradation temperature and activation energy of the PLA/ATH/OMMT nanocomposite determined by TGA are increased in comparison with the composition without addition of ATH and organoclay. The incorporation of OMMT in PLA/ATH results in further stabilization. Exfoliated and intercalated structures of clay in the matrix were observed by TEM and XRD.

V-0 rating of PLA composites was achieved, and the melt dripping was reduced during combustion. Cone calorimeter tests (samples of $100 \times 100 \times 6$ mm^3 and irradiance of 50 kW/m^2) showed that pHRR was strongly reduced for the introduction of 2 or 5 wt% in substitution of 40 wt% ATH (Fig. 1.16). Nevertheless, despite Time to Ignition was improved in comparison with pristine PLA for all compositions, it is lowered for the binary compositions ATH/OMMT in comparison with ATH alone.

The same authors associated ATH with a hyperbranched polymer (noted HBP6) with polyethylene glycol (PEG) chains, synthesized via bulk polymerization of diepoxide, primary amine, and monoepoxide [84].

It was found that a V-0 rating is obtained from a composition consisting of 30 wt% ATH in PLA. PLA/ATH/HBP6 composites, where the addition of HBP6 replaced a portion of the ATH in the range 1–4 wt%, exhibit higher LOI values than the sample without HBP6. Moreover, V-0 rating is obtained for all compositions, but total flaming combustion is lower in presence of HBP6.

Kiuchi et al. [85] incorporated higher amounts of ATH in PLA. 50 wt% of ATH containing a low concentration of alkali elements improved significantly the flame retardancy of PLA. The simultaneous addition of phenol resins (including phenol novolac resin and tri-phenol methane resin) and ATH improved the flame retardancy of the PLA composites. A high-molecular weight phenol novolac resin shows the highest flame-retarding effect. 5 wt% phenol resin added to 50 wt% ATH allows V-0 rating to be achieved, whereas only V-1 rating is obtained for 50 wt% ATH alone (Fig. 1.17). This enhancement effect on flame retardancy is ascribed by the authors to the formation of homogeneous char layers produced by the phenol resins on the surface of the PLA composites during combustion.

Fig. 1.17 UL 94 rating as a function of the percentage of ATH and addition of phenol novolac resin (PN resin), from Kiuchi et al. [85]. Total framing time represents the total burning time after ignitions at UL 94 test

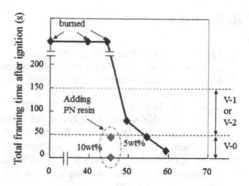

1.5   Conclusion

The routes to improve the fire reaction of biobased polymers, mainly biopolyesters and above all PLA, are not significantly different from the general strategies carried out for the most part of thermoplastic polymers. Even if it is difficult to make comparisons between the cone calorimeter results, due to various thicknesses of specimens tested and irradiances applied, the most efficient compositions correspond to the combination of intumescent systems, mainly based on inorganic phosphates, charring agents, possibly combined with different kind of nanoparticles. Furthermore, in order to reduce the environmental footprint of these materials, renewable components of such intumescent compositions have been particularly developed such as mainly lignins and starch. In parallel, novel phosphorous flame retardants were synthesized which allows to achieve a self-intumescent behaviour for some of them. The development of new kinds of functionalizations for engineering nanoparticles, such as carbon nanotubes, zirconium phosphate, layered double hydroxides for example, offers new opportunities to improve the fire reaction of biobased polymers among other polymers in which they can be incorporated. As for fossil based polyesters, polyamides or polyurethanes, reactive FRs were produced through the chemical modification of monomers, or by the incorporation of chemical groups in the backbone, mainly phosphorous ones. Nevertheless for biodegradable biobased polymers, the biodegradability of such modified components has to be investigated. Moreover, for the most part of solutions proposed to improve flame retardancy of biobased polymers, the influence of the FR systems proposed on the mechanical properties was not studied. This point is particularly critical for PLA which is rather brittle and exhibits a very low elongation at break. Hence, future challenges concern the development of FR systems devoted to a wider range of biobased polymeric materials as well as components able to less impair the functional properties as well as their biodegradability.

References

1. Shen L, Haufe J, Patel MK (2009) Product overview and market projection of emerging bio-based plastics. Utrecht University, www.epnoe.eu
2. http://www.succinity.com/images/succinity_broschure.pdf (2016) Final report FP 7 European project, grant agreement No. 289196
3. Alongi J, Han Z, Bourbigot S (2015) Intumescence: tradition versus novelty. A comprehensive review. Prog Polym Sci 51:28–73
4. Reti C, Casetta M, Duquesne S, Bourbigot S, Delobel R (2008) Flammability properties of intumescent PLA including starch and lignin. Polym Adv Technol 19:628–635
5. Wang X, Hu Y, Song L, Xuan S, Xing W, Bai Z, Lu H (2011) Flame retardancy and thermal degradation of intumescent flame retardant poly(lactic acid)/starch, biocomposites. Ind Eng Chem Res 50:713–720
6. Wu K, Hu Y, Song L, Lu HD, Wang ZZ (2009) Flame retardancy and thermal degradation of intumescent flame retardant starch-based biodegradable composites. Ind Eng Chem Res 48:3150–3157
7. Lyon RE, Walters RN (2004) Pyrolysis combustion flow calorimetry. J Anal Appl Pyrol 71:27–46
8. Cayla A, Rault F, Giraud S, Salaün F, Fierro V, Celzard A (2016) PLA with intumescent system containing lignin and ammonium polyphosphate for flame retardant textile. Polymers 8:331–346
9. Zhang R, Xiao X, Tai Q, Huang H, Yang J, Hu Y (2012) Preparation of lignin–silica hybrids and its application in intumescent flame-retardant poly(lactic acid) system. High Perform Polym 24:738–746
10. Zhang X, Xiao Q, Tai H, Huang J, Yang YHu (2013) The effect of different organic modified montmorillonites (OMMTs) on the thermal properties and flammability of PLA/MCAPP/lignin systems. J Appl Polym Sci 127:4967–4973
11. Morgan A, Wilkie CA (eds) (2010) Multicomponents FR systems: polymer nanocomposites combined with additional materials. In: Fire retardancy of polymeric materials. CRC Press (Chap. 12)
12. Fontaine G, Bourbigot S (2009) Intumescent polylactide: a nonflammable material. J Appl Polym Sci 113:3860–3865
13. Matusinovic Z, Wilkie CA (2012) Fire retardancy and morphology of layered double hydroxide nanocomposites: a review. J Mater Chem 22:18701–18704
14. Wang X, Zhou S, Xing WY, Yu B, Feng XM, Song L, Hu Y (2013) Self-assembly of Ni–Fe layered double hydroxide/graphene hybrids for reducing fire hazard in epoxy composites. J Mater Chem A 1:4383–4390
15. Dasari A, Yu ZZ, Cai GP, Mai YW (2013) Recent developments in the fire retardancy of polymeric materials. Prog Polym Sci 38:1357–1387
16. Wang DY, Leuteritz A, Wang Y-Z, Wagenknecht U, Heinrich G (2010) Preparation and burning behaviors of flame retarding biodegradable poly(lactic acid) nanocomposite based on zinc aluminum layered double hydroxide. Polym Deg Stab 95:2474–2480
17. Sue HJ, Gam KT (2004) Epoxy nanocomposites based on the synthetic α-zirconium phosphate layer structure. Chem Mater 16:242–249
18. Zhang R, Hu Y, Li BG, Chen ZY, Fan WC (2007) Studies on the preparation and structure of polyacrylamide/α-zirconium phosphate nanocomposites. J Mater Sci 42:5641–5646
19. Liu CH, Yang YJ (2009) Effects of α-zirconium phosphate aspect ratio on the properties of polyvinyl alcohol nanocomposites. Polym Test 28:801–807
20. Wang DY, Liu XQ, Wang JS, Wang YZ, Stec AA, Hull TR (2009) Preparation and characterization of a novel fire retardant PET/α-zirconium phosphate nanocomposite. Polym Degrad Stab 94:544–549

21. Liu XQ, Wang DY, Wang XL, Chen L, Wang YZ (2011) Synthesis of organo-modified α-zirconium phosphate and its effect on the flame retardancy of IFR poly(lactic acid) systems. Polym Deg Stab 96:771–777

22. Hu XP, Li WY, Wang YZ (2004) Synthesis and characterization of a novel nitrogen containing flame retardant. J Appl Polym Sci 94:1556–1561

23. Vahabi H, Ferry L, Longuet C, Otazaghine B, Negrell-Guirao C, David G, Lopez-Cuesta J-M (2012) Combination effect of polyhedral oligomeric silsesquioxane (POSS) and a phosphorus modified PMMA, flammability and thermal stability properties. Mater Chem Phys 136: 762–770

24. Didane N, Giraud S, Devaux E, Lemort G (2012) A comparative study of POSS as synergists with zinc phosphinates for PET fire retardancy. Polym Degrad Stab 97:383–391

25. Fox DM, Lee J, Citro CJ, Novy M (2013) Flame retarded poly(lactic acid) using POSS-modified cellulose. 1. Thermal and combustion properties of intumescing composites. Polym Degrad Stab 98:590–596

26. Fox DM, Novy M, Brown K, Zammarano M, Harris RH, Murariu M, McCarthy ED, Seppala JE, Gilman JW (2014) Flame retarded poly(lactic acid) using POSS-modified cellulose. 2. Effects of intumescing flame retardant formulations on polymer degradation and composite physical properties. Polym Degrad Stab 106:54–62

27. Gao L, Zheng G, Zhou Y, Hu L, Feng G, Zhang M (2014) Synergistic effect of expandable graphite, diethyl ethylphosphonate and organically-modified layered double hydroxide on flame retardancy and fire behavior of polyisocyanurate-polyurethane foam nanocomposite. Polym Degrad Stab 101:92–101

28. Murariu M, Dechief AL, Bonnaud L, Paint Y, Gallos A, Fontaine G, Bourbigot S, Dubois P (2010) The production and properties of polylactide composites filled with expanded graphite. Polym Degrad Stab 95:889–900

29. Zhu H, Zhu Q, Li J, Tao K, Xue L, Yan Q (2011) Synergistic effect between expandable graphite and ammonium polyphosphate on flame retarded polylactide. Polym Degrad Stab 96:183–189

30. Laachachi A, Cochez M, Leroy E, Gaudon P, Ferriol M, Lopez Cuesta JM (2006) Effect of Al₂O₃ and TiO₂ nanoparticles and APP on thermal stability and flame retardance of PMMA. Polym Adv Technol 17:327–334

31. Feng C, Liang M, Zhang Y, Jiang J, Huang J, Liu H (2016) Synergistic effect of lanthanum oxide on the flame retardant properties and mechanism of an intumescent flame retardant PLA composites. J Anal Appl Pyrol 122:241–248

32. Yang HE, Chapin JT, Gandhi P, Lackhouse T (2013) Micro-scale evaluation of flammability for cable materials. In: Proceeding of 62th international wire & cable symposium

33. Ke CH, Li J, Fang KY, Zhu Q-L, Zhu J, Yan Q, Wang YZ (2010) Synergistic effect between a novel hyperbranched charring agent and ammonium polyphosphate on the flame retardant and anti-dripping properties of polylactide. Polym Degrad Stab 95:763–770

34. Shabanian M, Kang NJ, Wang DY, Wagenknecht U, Heinrich G (2013) Synthesis of aromatic aliphatic polyamide acting as adjuvant in polylactic acid (PLA)/ammonium polyphosphate (APP) system. Polym Degrad Stab 98:1036–1042

35. Bocz K, Domonkos M, Igricz T, Kmetty Á, Bárány T, Marosi G (2015) Flame retarded self-reinforced poly(lactic acid) composites of outstanding impact resistance. Compos A 70:27–34

36. Carosio F, Laufer G, Alongi J, Camino G, Grunlan JA (2011) Layer-by-layer assembly of silica-based flame retardant thin film on PET fabric. Polym Degrad Stab 96:745–750

37. Garlotta DA (2001) A literature review of poly(lactic acid). J Polym Environ 9:63–84

38. Jing J, Zhang Y, Tang X, Zhou Y, Li X, Kandola BK, Fang Z (2017) Layer by layer deposition of polyethylenimine and bio-based polyphosphate on ammonium polyphosphate: A novel hybrid for simultaneously improving the flame retardancy and toughness of polylactic acid. Polymer 108:361–371

39. Levchik SV, Costa L, Camino G (1992) Effect of the fire-retardant, ammonium polyphosphate, on the thermal decomposition on of aliphatic polyamides. I. Polyamides 11 and 12. Polym Degrad Stab 36:31–41
40. Dorez G, Taguet A, Ferry L, Lopez-Cuesta JM (2013) Thermal and fire behavior of natural fibers/PBS biocomposites. Polym Degrad Stab 98:87–95
41. Dumazert L, Rasselet D, Pang B, Gallard B, Kennouche S, Lopez-Cuesta J-M. Thermal stability and fire reaction of poly(butylene succinate) nanocomposites using natural clays and FR additives. Polym Adv Technol (accepted)
42. Wang X, Yang H, Song L, Hu Y, Xing W, Lu H (2011) Morphology, mechanical and thermal properties of graphene-reinforced poly(butylene succinate) nanocomposites. Compos Sci Technol 72:1–6
43. Song L, Xuan S, Wang X, Hu Y (2012) Flame retardancy and thermal degradation behaviors of phosphate in combination with POSS in polylactide composites. Thermochim Acta 527:1–7
44. Murariu M, Dubois P (2016) PLA composites: from production to properties. Adv Drug Deliv Rev 107:17–46
45. Pack S, Bobo E, Muir N, Yang K, Swaraj S, Ade H, Cao C, Korach CS, Kashiwagi T, Rafailovich MH (2012) Engineering biodegradable polymer blends containing flame retardant-coated starch/nanoparticles. Polymer 53:4787–4799
46. Ju Y, Liao F, Dai X, Cao Y, Li J, Wang X (2016) Flame-retarded biocomposites of poly(lactic acid), distiller's dried grains with solubles and resorcinol di(phenyl phosphate). Compos A 81:52–60
47. Jing J, Zhang Y, Fang Z (2017) Diphenolic acid based biphosphate on the properties of polylactic acid: synthesis, fire behavior and flame retardant mechanism. Polymer 108:29–37
48. Chen X, Zhuo J, Jiao C (2012) Thermal degradation characteristics of flame retardant polylactide using TG-IR. Polym Degrad Stab 97:2143–2147
49. Laachachi A, Cochez M, Leroy E, Ferriol M, Lopez-Cuesta JM (2007) Fire retardant systems in poly(methyl methacrylate): interactions between metal oxide nanoparticles and phosphinates. Polym Degrad Stab 92:61–69
50. Braun U, Schartel B, Ficher MA, Jäger C (2007) Flame retardancy mechanisms of aluminium phosphinate in combination with melamine polyphosphate and zinc borate in glass-fibre reinforced polyamide 6,6. Polym Degrad Stab 92:1528–1545
51. Braun U, Schartel B (2008) Flame retardancy mechanisms of aluminium phosphinate in combination with melamine cyanurate in glass-fibre-reinforced poly(1,4-butylene terephthalate). Macromol Mater Eng 293:206–217
52. Bourbigot S, Fontaine G (2010) Flame retardancy of polylactide: an overview. Polym Chem 1:1413–1422
53. Isitman NA, Dogan M, Bayramli E, Kaynak C (2012) The role of nanoparticle geometry in flame retardancy of polylactide nanocomposites containing aluminium phosphinate. Polym Degrad Stab 97:1285–1296
54. Lin HJ, Liu SR, Han LJ, Wang XM, Bian YJ, Dong LS (2013) Effect of a phosphorus-containing oligomer on flame-retardant, rheological and mechanical properties of poly (lactic acid). Polym Degrad Stab 98:1389–1396
55. Avinc O, Day R, Carr C, Wilding M (2012) Effect of combined flame retardant, liquid repellent and softener finishes on poly(lactic acid) (PLA) fabric performance. Text Res J 82:975–984
56. Cheng XW, Guan JP, Tang RC, Liu KQ (2016) Improvement of flame retardancy of poly (lactic acid) nonwoven fabric with a phosphorus containing flame retardant. J Ind Text 46:914–928
57. Wei LL, Wang DY, Chen H-B, Chen L, Wang XL, Wang YZ (2011) Effect of a phosphorus-containing flame retardant on the thermal properties and ease of ignition of poly (lactic acid). Polym Degrad Stab 96:1557–1561
58. Wang DY, Song YP, Lin L, Wang XL, Wang YZ (2011) A novel phosphorus-containing poly (lactic acid) toward its flame retardation. Polymer 52:233–238

59. Yuan XY, Wang DY, Chen L, Wang XL, Wang YZ (2011) Inherent flame retardation of bio-based poly(lactic acid) by incorporating phosphorus linked pendent group into the backbone. Polym Degrad Stab 96:1669–1675

60. Zhan J, Song L, Nie S, Hua Y (2009) Combustion properties and thermal degradation behavior of polylactide with an effective intumescent flame retardant. Polym Degrad Stab 94:291–296

61. Zhao X, Gao S, Liu G (2016) A THEIC-based polyphosphate melamine intumescent flame retardant and its flame retardancy properties for polylactide. J Anal Appl Pyrol 122:24–34

62. Liao F, Ju Y, Dai X, Cao Y, Li J, Wang X (2015) A novel efficient polymeric flame retardant for poly (lactic acid) (PLA): synthesis and its effects on flame retardancy and crystallization of PLA. Polym Degrad Stab 120:251–261

63. Li Z, Wei P, Yang Y, Yan Y, Shi D (2014) Synthesis of a hyperbranched poly(phosphamide ester) oligomer and its high-effective flame retardancy and accelerated nucleation effect in polylactide composites. Polym Degrad Stab 110:104–112

64. Tao K, Li J, Xu L, Zhao X, Xue L, Fan X, Yan Q (2011) A novel phosphazene cyclomatrix network polymer: design, synthesis and application in flame retardant polylactide. Polym Degrad Stab 96:1248–1254

65. Tang G, Wang X, Xing W, Zhang P, Wang B, Hong N, Yang W, Hu Y, Song L (2012) Thermal degradation and flame retardance of biobased polylactide composites based on aluminum hypophosphite. Ind Eng Chem Res 51:12009–12016

66. Tang G, Wang X, Zhang R, Wang B, Hong N, Hu Y, Song L, Gong X (2013) Effect of rare earth hypophosphite salts on the fire performance of biobased polylactide composites. Ind Eng Chem Res 52:7362–7372

67. Gallo E, Schartel B, Acierno D, Russo P (2011) Flame retardant biocomposites: synergism between phosphinate and nanometric metal oxides. Eur Polym J 47:1390–1401

68. Bocz K, Szolnoki B, Wladyka-Przybylak M, Bujnowicz K, Harakaly G, Bodzay B (2013) Flame retardancy of biocomposites based on thermoplastic starch. Polimery 58:385–394

69. Negrell C, Frenehard O, Sonnier R, Dumazert L, Briffaud T, Flat JJ (2016) Self-extinguishing bio-based polyamides. Polym Degrad Stab 134:10–18

70. Lligadas G, Ronda JC, Galia M, Cadiz V (2006) Synthesis and Properties of thermosetting polymers from a phosphorus containing fatty acids derivative. J Polym Sci Part A: Polym Chem 44:5630–5644

71. Lligadas G, Ronda JC, Galia M, Cadiz V (2006) Synthesis and Properties of thermosetting polymers from a phosphorus containing fatty acids derivative. J Polym Sci Part A: Polym Chem 44:6717–6727

72. Montero de Espinoza L, Ronda JC, Galia M, Cadiz V (2009) A straightforward strategies for the efficient synthesis of acrylate and phosphine oxide-containing vegetable oils and their crosslinked materials. J Polym Sci Part A Polym Chem 47:4051–4063

73. Zhang L, Zhang M, Zhou Y, Hu L (2013) The study of mechanical behavior and flame retardancy of castor oil phosphate-based rigid polyurethane foam composites containing expanded graphite and triethyl phosphate. Polym Degrad Stab 98:2784–2794

74. Liu XQ, Wang DY, Wang XL, Chen L, Wang YZ (2013) Synthesis of functionalized α-zirconium phosphate modified with intumescent flame retardant and its application in poly (lactic acid). Polym Degrad Stab 98:1731–1737

75. Ding P, Kang B, Zhang J, Yang J, Song N, Tang S, Shi L (2015) Phosphorus-containing flame retardant modified layered double hydroxides and their applications on polylactide film with good transparency. J Coll Interf Sci 440:46–52

76. Hu Y, Xu P, Gui H, Wang X, Ding Y (2015) Effect of imidazolium phosphate and multiwalled carbon nanotubes on thermal stability and flame retardancy of polylactide. Compos A 77:147–153

77. Costes L, Laoutid F, Aguedo M, Richel A, Brohez S, Delvosalle C, Dubois P (2016) Phosphorus and nitrogen derivatization as efficient route for improvement of lignin flame retardant action in PLA. Eur Polym J 84:652–667

78. Costes L, Laoutid F, Khelifa F, Rose G, Brohez S, Delvosalle C, Dubois P (2016) Cellulose/phosphorus combinations for sustainable fire retarded polylactide. Eur Polym J 74:218–228
79. Costes L, Laoutid F, Dumazert L, Lopez-Cuesta JM, Brohez S, Delvosalle C, Dubois P (2015) Metallic phytates as efficient bio-based phosphorous flame retardant additives for poly (lactic acid). Polym Degrad Stab 119:217–227
80. González A, Dasari A, Herrero B, Plancher E, Santarén J, Esteban A, Lim SH (2012) Fire retardancy behavior of PLA based nanocomposites. Polym Degrad Stab 97:248–256
81. Dhanushka Hapuarachchi T, Peijs T (2010) Multiwalled carbon nanotubes and sepiolite nanoclays as flame retardants for polylactide and its natural fibre reinforced composites. Compos A 41:954–963
82. Ferry L, Gaudon P, Leroy E, Lopez-Cuesta JM (2005) Fire retardancy of polymers: new applications of mineral fillers. In: Le Bras M, Wilkie C, Bourbigot S, Duquesne S, Jama C (eds) Intumescence in ethylene-vinyl acetate copolymer filled with magnesium hydroxide and organoclays. The Royal Society of Chemistry, Oxford, pp 302–312 (chapter 22)
83. Cheng KC, Yu C-B, Guo W, Wang SF, Chuang TH, Lin Y-H (2012) Thermal properties and flammability of polylactide nanocomposites with aluminum trihydrate and organoclay. Carbohydr Polym 87:1119–1123
84. Cheng KC, Chang SC, Lin YH, Wang CC (2015) Mechanical and flame retardant properties of polylactide composites with hyperbranched polymers. Compos Sci Technol 118:186–192
85. Kiuchi Y, Iji M, Yanagisawa T, Shukichi T (2014) Flame-retarding polylactic-acid composite formed by dual use of aluminum hydroxide and phenol resin. Polym Degrad Stab 109: 336–342

Chapter 2
Biobased Flame Retardants

2.1 Issues and Objectives

A great part of commercially available flame retardants are oil-derived organic compounds (e.g. organo-halogenated, organo-phosphorous, organo-nitrogen compounds). As part of oil-based products, they face the same issues: growing scarcity of petroleum, geopolitical problems, impact on global warning. Moreover some of these compounds (i.e. the halogenated compounds) have received a bad press because there are suspected to provoke specific health and environment concerns. In the prospect to propose more environmentally friendly materials, it is necessary to go further in the application of the Green Chemistry principles as defined by Anastas and Warner [1] and thus to promote the development of biobased additives for polymers [2]. Compared to other additives, it must be said that flame retardants were not pioneers in that area. Indeed, most of the studies that will be presented in this chapter have been carried out for less than five years. Nevertheless times are changing and in the recent period the scientific community has produced a significant number of papers dedicated to the development of flame retardants from renewable resources. The ban of some halogenated compounds and the search for alternative solutions has been definitely an important driving force for the development of such new biobased additives. But another challenging insight is also to take advantage of the particular chemical structure of bio-resources to promote and emphasize specific flame retardant mechanisms.

This chapter will first describe the different biomolecules likely to be used as starting matter for flame retardants, highlighting their chemical structure and their intrinsic thermal properties. Then a review of biobased flame retardant systems studied in the literature will be proposed. Systems have been classified according to their expected mode of action. Finally, this chapter will close off with some thoughts about future industrial developments of biobased flame retardants.

© The Author(s) 2018
R. Sonnier et al., *Towards Bio-based Flame Retardant Polymers*,
Biobased Polymers, DOI 10.1007/978-3-319-67083-6_2

2.2   Thermal Behavior of Biomass-Based Matter

Biobased compounds generally refer to compounds that can be obtained or derived from biomass which is defined as the biological matter that can be found on earth. It includes plants, animals and microorganisms. Apart from bacteria, the total biomass on land represents about 560 billion tons of carbon. Vassilev et al. [3] gave an overview of the chemical composition of biomass. Their analysis relies on the general classification of biomass in groups, sub-groups, varieties and species. Even if this composition is very variable depending on the group, they were able to identify the most abundant elements which in decreasing order are C, H, N, Ca, K, Si, Mg, Al, S, Fe, P, Cl, Na, Mn, and Ti. It is worthy to note that some of these elements are known to have a flame retardant effect.

In order to be used as raw matter for fuels, heat, power or chemicals, biomass has to be converted using various processes. This set of processes is the so-called biorefinery, a concept developed by analogy with the oil industry [4]. The purpose of a biorefinery is to extract, separate and modify the different biochemical compounds of biomass and to add value to these products and their intermediates. Therefore a better way of imaging how biomass can be used to develop flame retardants consists in considering its biochemical composition. Thus, four main families of compounds can be distinguished: carbohydrates, proteins, lipids and phenolic compounds (Fig. 2.1). These biomolecules can be used as such or

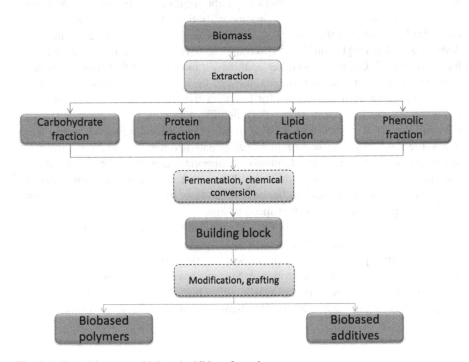

Fig. 2.1   From biomass to biobased additives for polymers

converted to derivatives by chemical or biological treatments. Thus biobased building blocks can be obtained and further modified to bring flame retardant functionalities.

2.2.1 Carbohydrates

Carbohydrates, also known in biochemistry as saccharides, are biological molecules containing carbon, hydrogen and oxygen atoms. They include low molecular weight compounds referred to as sugars (e.g. glucose, lactose) or more complex molecules such as oligo and polysaccharides. The compounds that got the more attention for flame retardant development were probably polysaccharides, and more particularly cellulose, starch and chitosan.

Cellulose is the main structural component of vegetal cell wall in plants. It is the most abundant source of organic matter on earth since it is estimated that plants synthesize 50 to 100 billion tons of cellulose each year. Cellulose is a linear homopolymer composed of D-glucose units linked by β 1-4 glycosidic bonds [5] (see Fig. 2.2). The degree of polymerization of native cellulose varies according to the source from 1000 to 30,000. The individual polysaccharide chains are bound together by hydrogen bonds into a fibrous structure called microfibrils whose diameter is about 25 nm. Parts of microfibrils called micelles are arranged in an orderly manner and give to cellulose its crystalline properties. Microfibrils are bundled together into fine threads that coil around one another to form macrofibrils. At a higher scale, arrangement of macrofibrils forms cellulose fibers that act as reinforcement in plant cell walls.

The thermal decomposition of cellulose occurs in one main step between 300 and 400 °C, the higher mass loss rate being observed around 370 °C (Fig. 2.3). Below 300 °C cellulose undergoes dehydration reactions that give rise to a small mass loss. Dehydration reactions may be intermolecular or intramolecular thus creating crosslinking and double bonds. This leads to the formation of an intermediary compound called active cellulose or anhydrocellulose. Above 300 °C the main degradation step corresponds to depolymerization that occurs via the breaking of the glycosidic linkage (transglycosylation). The decomposition of cellulose gives various kinds of anhydro-saccharides the main one being levoglucosan (1,6-anhydro-β-D-glucopyranose) that may represent up to 60 wt% of volatile yield. During this step, the amount of furan and benzene rings increases progressively in the condensed phase while aliphatic groups disappear. At high temperature (800 °C), a stable char representing 15 wt% of the initial mass is observed. This residue has a structure similar to that of lignin char. It should be added that the degradation pathway of cellulose may be modified depending on the heating conditions and the presence of other components in the surrounding. Thus at low heating rate, dehydration reactions are favored and the char yield is enhanced [6]. Similarly it was highlighted by Dorez et al. [7] that the presence of a low amount of lignin in flax fiber was likely to change the degradation pathway of cellulose due to

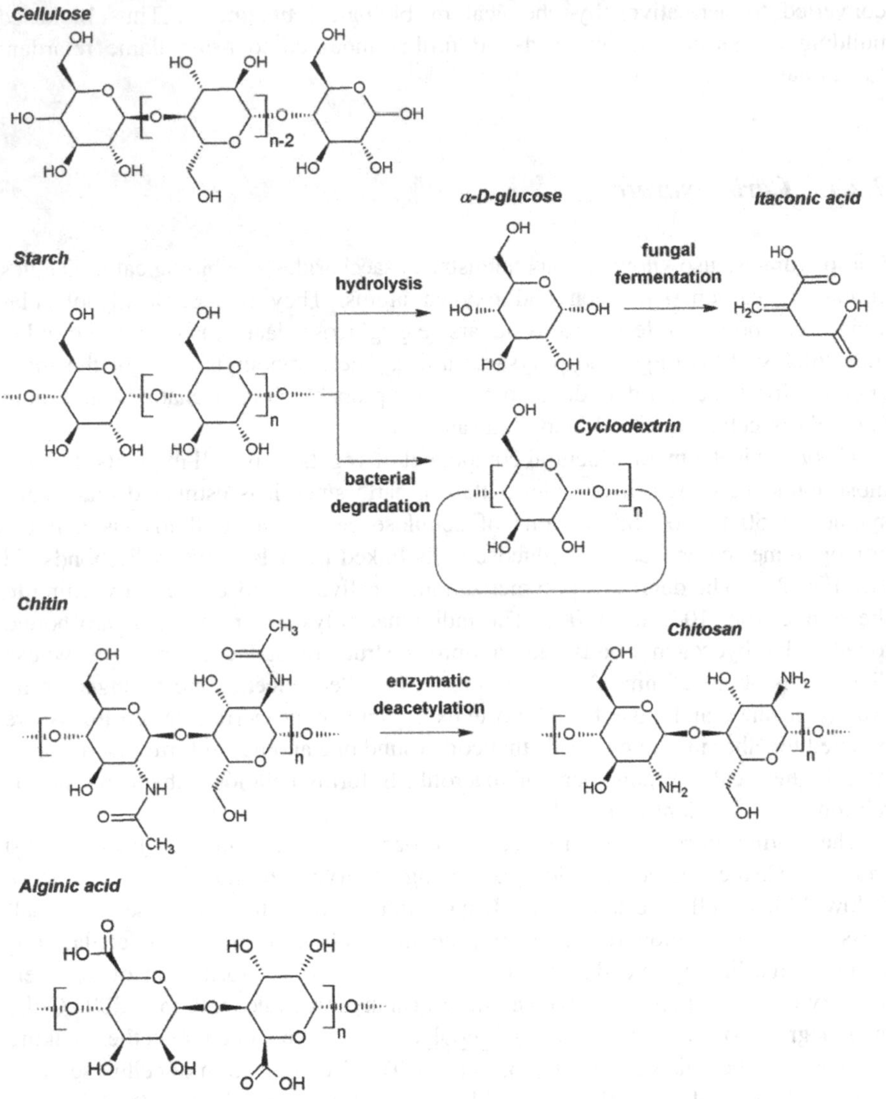

Fig. 2.2 Examples of carbohydrates used in flame retardant systems

the release of acidic compounds that promote dehydration and hence charring at the expense of depolymerization.

Starch is a polymer that is synthesized by plants as energy storage [8]. It takes the form of granules measuring 1–200 μm in diameter. Starch is very similar to cellulose in its chemical formula since it is also composed of D-glucose units. Actually starch contains two types of macromolecules: amylose which is a linear polymer constituted by glucose units bonded by a α 1-4 linkage and amylopectin

Fig. 2.3 Thermogravimetric curves of some carbohydrates

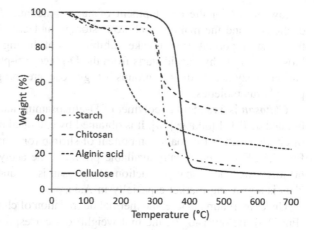

which is a branched polymer where branching occurs through α 1-6 linkage (see Fig. 2.2). Amylose whose fraction represents 15–25 wt% in starch exhibits molecular ranging from 0.2 to 2×10^6 g/mol while amylopectin which represents 75–85 wt% in starch has higher molecular weight ranging from 100 to 400×10^6 g/mol. Approximately 70 million tons of starch are produced per year world-wide.

The thermogravimetric curve of starch under inert atmosphere highlights two main weight losses (Fig. 2.3). The first weight loss occurring at low temperature up to 110 °C corresponds to evaporation/physical dehydration phenomena. As a matter of fact, starch is a highly hygroscopic material [9]. The second mass loss corresponds to chemical dehydration and thermal decomposition of starch and starts from 300 °C. In this range of temperature condensation between hydroxyl groups is observed leading to the formation of ether bonds. Condensation inside the glucose ring may also occur, leading to its breakdown. C–C bonds as well as aldehyde groups are formed. Above 500 °C carbonization occurs, leading to an increase of aromatic carbons at the expense of aliphatic carbons. The thermal decomposition of starch gives a 15–20 wt% residue at high temperature.

Starch industry not only includes the extraction and refinement of starches but also the production of starch derivatives. Starch can be hydrolyzed into simpler carbohydrates also called dextrins by acids or enzymes. As an example, *cyclodextrin* that will be cited later in the text as component of flame retardant systems was shown to be obtained via the degradation of starch by the bacillus amylobacter bacteria (see Fig. 2.2). Cyclodextrin is a cage molecule that may encapsulate other compounds and therefore appears as very attractive for the food and pharmaceutical sectors. Other polyols such as sorbitol or *isosorbide* may also be relevant to be used in flame retardant formulations. Fermentation is another process that enables to convert starch into derivatives. For example the use of fungi such as Aspergillus itaconicus or Aspergillus terreus allows for the conversion of glucose or molasses into *itaconic acid* (see Fig. 2.2). The fermentation of grape stock to make wine generates *tartaric acid*. These two organic acids have been used as raw matter for the development of flame retardant. *Phytic acid* is a saturated cyclic acid and the main phosphorous storage molecule found in plant seeds. It is the most important iron uptake inhibitor contributing to iron deficiency. The biosynthesis of phytic acid starts from the D-glucose-6-phosphate. By the action of various enzymes hydroxyl groups of glucose are progressively substituted by phosphorus moieties.

Chitosan is a random copolymer of D-glucosamine and N-acetyl-D-glucosamine bonded by β 1-4 linkage [10]. It is obtained by chemical or enzymatic deacetylation of chitin which is the main component of shrimp (or other crustaceans) shells (see Fig. 2.2). For commercially available products, the acetylation degree varies from 60 to 100%. The global production of chitosan is around 20×10^3 tons per year. This is a growing market especially in Asia.

Under inert atmosphere, the thermal degradation of chitosan occurs in three steps (Fig. 2.3). Below 140 °C, the first weight loss corresponds to the release of water loosely bonded [11]. The second and main step of degradation occurring between 250 and 350 °C is assigned to further dehydration, deacetylation and depolymerization of chitosan. Above 400 °C a low mass loss rate attributed to residual decomposition reaction is observed. At 500 °C the char is as high as 40 wt%. As shown by Moussout et al. [12], this char is much lower (circa 20 wt%) when degradation occurs under air atmosphere. By comparing chitin and chitosan, these authors also evidenced that deacetylation degree modified both the hygroscopicity and thermal stability of the carbohydrate.

Alginates refer to the derivatives of alginic acid and alginic acid itself [13]. These are carbohydrates present in the cell wall of brown algae as calcium, magnesium and sodium salts of alginic acid. Alginic acid is a copolymer of mannuronic acid and guluronic acid, repeating units being bonded by a β 1-4 linkage (see Fig. 2.2). The proportion and distribution of comonomers are determining parameters to explain the physical and chemical properties of the polymer. Thus alginates could be considered as an anionic polysaccharide.

The thermal behavior of alginic acid and sodium alginate was studied by Soares et al. [14]. Under nitrogen, two main steps of degradation are highlighted (Fig. 2.3). At low temperature alginates exhibit a dehydration process. The moisture content was found to be higher for sodium alginate (15 wt%) than for alginic acid (10 wt%). The main decomposition step occurs between 150 and 300 °C. In the case of alginic acid, it leads to carbonaceous residue that was found to be between 21 and 26 wt% at 800 °C depending on the authors [15, 16]. In the case of sodium alginate, Soares et al. obtained a 22 wt% residue consisting mainly of sodium carbonate.

2.2.2 Proteins

Proteins are linear biological polymers found in living cells [17]. They are made of sequence of amino acids bonded by peptidic linkage (Fig. 2.4). There are only 20

naturally occurring amino acids that can be classified according to the polarity and
the charge of the side chain that they bear. Proteins exhibit a very complex structure
that can be described through 4 levels of organization. The primary structure corresponds to the sequence of amino acids in the macromolecule. The secondary
structure corresponds to the local structure of the protein chain stabilized by
hydrogen bonds. Hydrogen bonds are related to the presence of carbonyl and amine
groups. It leads to structures such α-helix or β-sheet. The tertiary structure describes
the folding of a single protein and defines its overall shape that is stabilized by
different kinds of interactions. Globular, fibrous and membrane proteins are the
most encountered tertiary structures. Finally the quaternary structure describes the
structure formed by several proteins.

Proteins may play various roles in living cells. Enzymes which are a large family
of proteins act as catalyst for biochemical reactions. Other proteins are involved as
sensor in the communication system that governs the cell activities. They may also
participate to the transport of smaller molecules. Finally, some polypeptides such as
fibrous proteins play a structural role and bring stiffness to cell components.

It must be underlined that some proteins involved in signal transduction may
undergo phosphorylation and dephosphorylation modification. Well known
examples of phosphoproteins are casein (protein of milk) and phosvitine (protein of
egg). The amino acids residues that may be modified by phosphate groups are
usually serine, threonine or tyrosine (Fig. 2.5).

The thermal stability and the pyrolysis products of proteins tightly depend on
their amino-acid composition [18]. As an example (Fig. 2.6), the thermal degradation of casein was studied by Mocanu et al. [19]. Their results indicate that casein
undergoes a small weight loss below 170 °C that was assigned to the release of
water physically retained within the proteins. Under nitrogen, the main degradation
step occurs between 250 and 380 °C with the release of CO_2, CO, water, ammonia
and isocyanic acid. The decomposition lets a thermally stable residue that can be as
high as 25 wt% at 600 °C.

The primary structure of proteins i.e. the sequence of amino acids is given by the
genetic information carried by deoxyribonucleic acid (DNA). Therefore it can be

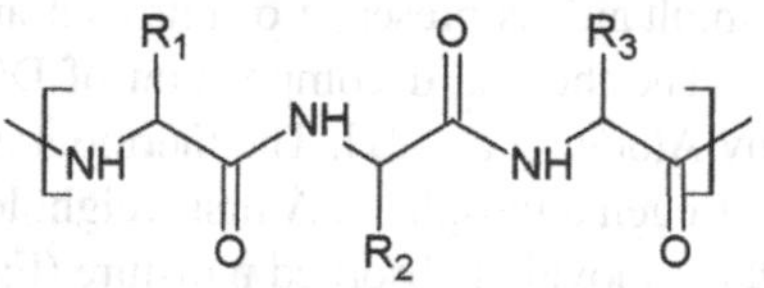

Fig. 2.4 Chemical structure
of proteins

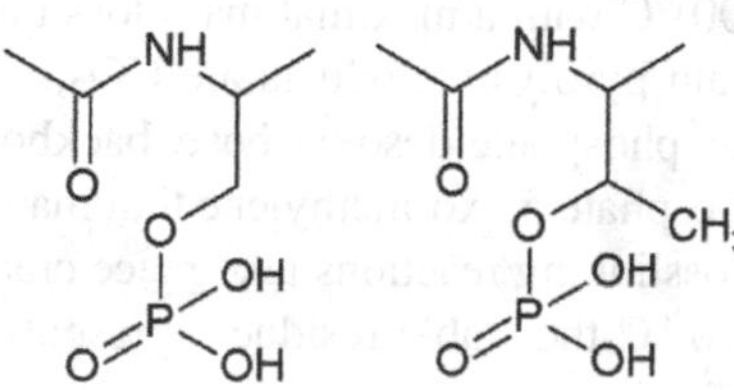

Fig. 2.5 Examples of
phosphoprotein residues
(phosphoserine and
phosphothreonine)

Fig. 2.6 Thermogravimetric curves of casein and DNA from [19, 21]

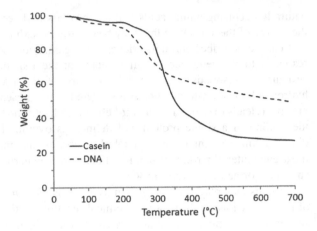

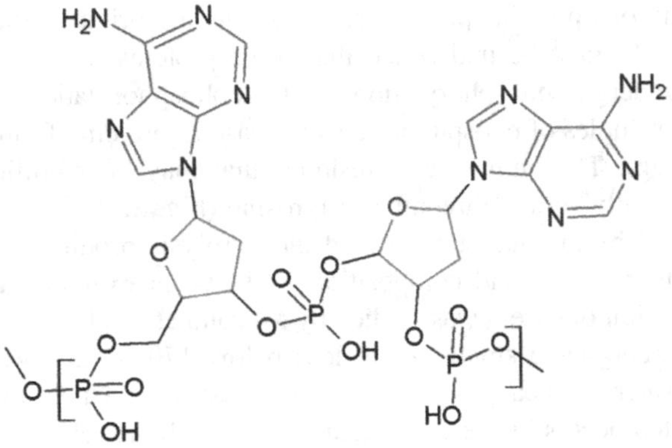

Fig. 2.7 Structure of DNA including sugar, phosphate and base units

said that proteins and DNA are closely tied in biosynthesis. As shown in Fig. 2.7, DNA exhibits a complex structure consisting of a phosphate-desoxyribose backbone with a base (adenine, cytosine, guanine or thymine) attached to the sugar ring [20]. This molecule has raised interest in the field or flame retardancy due to the simultaneous presence of nitrogen and phosphorus.

The thermal decomposition of DNA from herring sperm was studied in details by Alongi et al. [21]. The thermogravimetric curves are relatively similar in air and nitrogen atmosphere. A first weight loss is observed below 150 °C corresponding to the removal of absorbed moisture (Fig. 2.6). Some volatiles (e.g. isocyanic acid) are emitted around 160 °C however the main decomposition occurs between 200 and 400 °C with a maximal mass loss rate between 230 and 260 °C. Above 200 °C the main pyrolysis products are CO_2, water and ammonia. At still higher temperature the phosphate-desoxyribose backbone degrades producing 2,3 dihydrofuran 3,5 phosphate 5 exo-methylene that may further dimerize or polymerize. Above 400 °C crosslinking reactions take place creating a thermally stable P-N based structure. At 700 °C the stable residue represents circa 50% of the initial mass of DNA.

2.2.3 Lipids

Lipids are a group of naturally occurring molecules whose main biological functions are energy storing, signal transmission and structural material for cell walls. They are generally considered as hydrophobic or amphiphilic molecules indicating that at least one part of these molecules is non polar [22]. Taking into account the diversity of their chemical structure, there is no universal classification of lipids. In 2005 Fahy et al. [23] proposed a comprehensive classification of lipids in accordance with IUPAC rules including eight categories: fatty acids (e.g. ricinoleic acid), glycerol-lipids, glycerol-phospholipids (e.g. phosphatidylethanolamine), sphingolipids, sterol lipids, prenol lipids, saccharolipids, polyketides. To these true lipids, other compounds with a lipid character can be added: phenolic lipids (e.g. cardanol), terpenes (e.g. saponin), steroids. Some of the above-mentioned lipids are presented in Fig. 2.8.

The thermal stability and the decomposition mechanisms of lipids tightly depend on their nature. Thus, no general scheme can be described. However the thermal analysis of some vegetable oils (sunflower, soybean, castor, jojoba) indicates that the decomposition occurs generally in the range 300–550 °C and leaves no residue at high temperature [24–26]. Considering the presence of hydrocarbon chains in their structure, the combustion of lipids is likely to be highly exothermic. Therefore lipids appear not to be the best candidates for being the basis of biobased flame retardants.

Nevertheless, for many years, plant oils and their derivatives have been used as building blocks in polymer chemistry with the aim to ensure a more sustainable production of macromolecular materials [27]. Due to the presence of various functional groups (hydroxyls, carboxylic acids, double bonds), a great variety of reactions may be envisaged. Hence some lipid derivatives such as undecylenic acid or cardanol have been used to prepare phosphorus containing monomers that can act as reactive flame retardant, especially in thermosets.

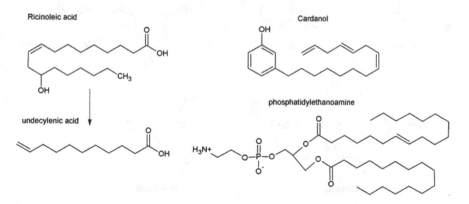

Fig. 2.8 Examples of lipids that may be used for biobased flame retardant synthesis

Besides, among the above-mentioned list of lipids, some may retain attention for prospective works owing to their chemical composition. Phospholipids and sphingolipids are two categories of lipids containing phosphorus and nitrogen atoms in their backbone. Considering the influence of these atoms in the fire behavior of polymers, these lipids may represent potential raw matter for new biobased flame retardants.

2.2.4 Phenolic Compounds

Natural polyphenols are a class of organic molecules found in plants and characterized by the presence of phenolic units bearing one or several hydroxyl groups associated in a high molecular weight complex structure. They result from the secondary metabolism of plants through the shikimate pathway [28]. They range from simple molecules such as hydroquinone to high polymerized structures such as tannins. Polyphenols can be classified according to the sequences in their carbon backbone. Thus simple phenols correspond to a C_6 sequence (e.g. phloroglucinol), phenolic acids correspond to a C_6–C_1 sequence, flavonoids correspond to a C_6–C_3–C_6 sequence (e.g. catechin), lignins correspond to $(C_6$–$C_3)_n$ sequences and condensed tannins correspond to $(C_6$–C_3–$C_6)_n$ sequences. The molecules that have raised the interest of researchers as raw matter for flame retardants belong mainly to simple phenols such as phloroglucinol or eugenol (Fig. 2.9) and to polymerized structures (tannins and lignins).

Lignin is the second most abundant natural polymer after cellulose and the first aromatic polymer. Lignins are mainly found in higher plants and in some algae.

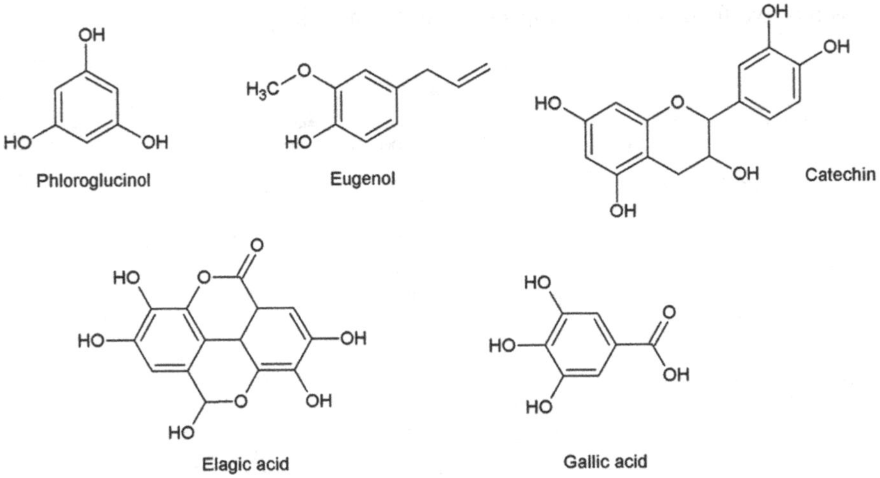

Fig. 2.9 Phenolic compounds

Fig. 2.10 TGA curves of lignin, pine tannins and catechin from [7, 30]

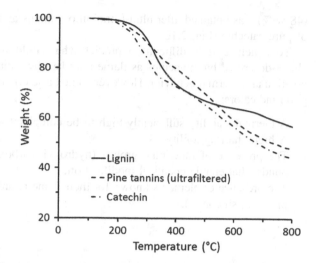

They play an important role in the mechanical properties of cell walls by bringing rigidity and in the protection against decomposition by giving waterproofness. On a chemical point of view lignins can be viewed as polymers of monolignols. They are based on three main monomers: coumarylic acid, coniferylic acid and sinapylic acid. The fraction of each monomer varies depending on the plant, the species, the organ and the tissue.

The thermal decomposition of lignin has been reviewed by Brebu and Vasile [29]. In comparison with other components of biomass, lignin degrades over a broader range of temperature between 200 and 500 °C (Fig. 2.10). The thermo-gravimetric curve generally shows a first weight loss between 100 and 180 °C that corresponds to the release of water physically bonded into the raw matter. Strictly speaking, the decomposition starts from circa 200 °C. In a first step (200–260 °C), low molecular weight products resulting from the propanoid side chain cleavage are released. Then the main degradation step (275–450 °C) corresponds to the cleavage of the main chain either by C–C and β-scission or by aryl-ether cleavage. At this point, a large quantity of methane is evolved. Above 500 °C, further rearrangements and condensation of the aromatic structure occur, leading to the formation of a significant char yield (57 wt% at 600 °C) and the release of dihydrogen in the gas phase.

Tannins are phenolic biomolecules that can be found in different parts of plants (bark, leaves, roots, leaves, fruits). They play a major role in the defense against pests. Two types of tannins can be distinguished: hydrolysable tannins which are based on gallic or ellagic acids and condensed tannins that are based on catechin or flavanol (Fig. 2.9). The thermal degradation of tannins extracted from pine bark was studied by Gaugler and Grigsby [30]. It was highlighted that the decomposition occurs over a wide range of temperature (180–800 °C) and was dependent on the mode of extraction and the degree of purification. Whatever the process, a high amount of char was formed at high temperature (>30 wt%). The largest char residue

(48 wt%) was obtained after ultra-filtration of tannins and was quite similar to that of pure catechin (Fig. 2.10).

To conclude it is difficult to predict what could be the interest of all the above-described biomolecules as flame retardant by simply examining their composition and thermal behavior. However some basic criteria may be supposed to be good indicators:

- A thermal stability sufficiently high to be compliant with polymer processing,
- A high charring ability,
- The presence of functional groups (hydroxyl, carboxylic acid, amine, double bonds) that enable chemical modification,
- The presence of elements known for their flame retardant activity (phosphorus, nitrogen, silicon…).

2.3 Strategies for Flame Retardancy with Biobased Compounds

The main strategies used for achieving flame retardancy of polymers have been reviewed in different papers [31, 32]. The modes of action of flame retardants are generally classified according to the nature of the involved mechanisms (physical or chemical action) and according to the location where these mechanisms take place (gas phase or condensed phase). Regarding all the biobased flame retardant systems that will be reported in the following, it is noteworthy that the improvement of the fire behavior was essentially achieved through one preferential mode of action the so-called charring effect. It means that the intended effect was to create at the sample surface and during combustion a charred layer. This char generates a double positive influence: (i) on one side a part of carbon atoms constituting the polymer structure is fixed within the residue thus decreasing the amount of volatile fuel and therefore the amount of heat released by the combustion reaction; (ii) on the other side the charred layer acts as a protective shield that slows down the heat transfer to the underlying polymer and modifies the kinetics of fuel diffusion to the flame. The effect of charring may be further enhanced if an expanded layer is formed; this is the so-called intumescence phenomenon. In this case it is necessary that a gas acting as blowing agent was released simultaneously with the charring process.

In the following the works will be presented according to the way renewable resources were used as flame retardants systems. It means (i) as unique component of flame retardant system, (ii) in combination with traditional phosphorus or nitrogen flame retardant, (iii) after chemical modification, (iv) after chemical modification and insertion in the polymer chain.

2.3.1 Bio-resources as Intrinsic Flame Retardant

2.3.1.1 Lignin

Among the renewable resources described in the preceding paragraph, lignin was probably the one giving the highest char yield at high temperature. That is surely the reason why lignin was tested as single additive to improve the fire behavior of polymers. Gallina et al. [33] studied the burning behavior of a polypropylene (PP) containing 20 wt% of lignin. Under a 25 kW/m^2 heat flux with cone calorimeter, this blend exhibits a peak of heat release rate (pHRR) three times lower than the pure PP (Fig. 2.11). The result was even better than those obtained with more classical flame retardant systems e.g. melamine or boric acid. This decrease was achieved with a final residue of only 6 wt%. The sole negative point was a decrease of time-to-ignition (Fig. 2.12) but it was also the case for nitrogen or phosphorus containing flame retardant. In a second paper [34], the same group obtained still interesting fire performance by decreasing the lignin content down to 15 wt% in PP. The reduction of pHRR was significant, moreover very low smoke opacity and CO yield were measured.

In a later paper, Song et al. [35] investigated the effect of reactive compatibilization on the thermal and fire properties of acrylonitrile-butadiene-styrene (ABS)/lignin blends. Similarly to what was observed in PP, it was evidenced that lignin due to its lower thermal stability induces a slight decrease of the onset of degradation. In return, increasing the lignin content enables slowing down the degradation process and increasing the char residue. In cone calorimeter test, the presence of lignin leads to a 32% decrease of pHRR (Fig. 2.11). It was highlighted that compatibilization plays a positive role by reducing the mass loss rate and increasing the char residue. This was attributed to the cohesivity of the residue which is enhanced when lignin is well dispersed in the polymer matrix.

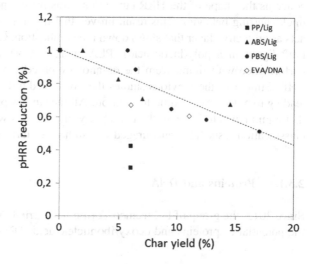

Fig. 2.11 Relative variation of pHRR versus char yield polymers containing biobased intrinsic flame retardant from [33, 34, 36, 38]

Fig. 2.12 Relative variation
of time to ignition versus
additive content of polymers
containing biobased intrinsic
flame retardant from [33, 34,
36, 38]

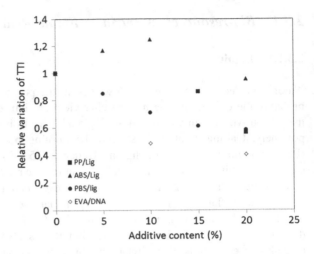

More recently Ferry et al. [36] compared the thermal and fire properties of two types of lignin differing by their extraction process (respectively kraft and organosolv process). The two lignins exhibit very similar thermogravimetric curves even if alkali lignin (kraft) gives a higher residue. More significant differences were evidenced when measuring the heat release rate during combustion either at small scale with micro-calorimeter of combustion (PCFC) or at bench scale with cone calorimeter. Alkali lignin shows lower pHRR and THR. This was assigned to the presence of sulfur resulting from the kraft process. It was assumed that sulfur containing moieties decomposed during lignin pyrolysis by releasing sulfur dioxide thus limiting the release of heat. The release of SO_2 was confirmed by using a FTIR/cone calorimeter coupling. Used at a 20 wt% content, alkali lignin induces an increase in polybutylene succinate (PBS) ignitability (Fig. 2.12) but also a signif-icant decrease of pHRR (49%) as shown in Fig. 2.11. The most significant change concerns the shape of the HRR curve that turns from a non-charring behavior to a thick charring behavior. This result proves that lignin promotes a char residue that acts as protective layer that slows down the combustion. Comparing organosolv and kraft lignin in a poly(lactic acid) (PLA) matrix, Costes et al. [37] found better performance with lignin from the organosolv process, with a strong reduction of THR. Similar to the previous studies they observed a decrease of thermal stability leading to a decrease of time-to-ignition. All the studies presented here-above show that lignin tends to increase the ignitability of polymers what is penalizing in many classification tests. This encouraged the authors to turn to lignin modification.

2.3.1.2 Proteins and DNA

Since 2013, the group of Politecnico Torino has carried out several studies to assess the potential of proteins and deoxyribonucleic acid (DNA) as flame retardant [39].

There were several motivations for using such biomacromolecules: firstly some of them can be considered as waste or by-products from food industry and therefore are low cost raw matter; secondly DNA despite its current high cost might find prospective applications due to the development of large scale extraction methods; finally these biomacromolecules present the advantage to contain nitrogen and phosphorus, two elements of great interest for flame retardancy.

The main field of applications of proteins and DNA was the treatment of textiles and foams due to the ability of these macromolecules to make films [40]. Bosco et al. [41] studied whey protein isolate used as coating onto cotton fabrics. Whey proteins are polypeptides exhibiting a globular tertiary structure. They represent circa 20 wt% of proteins in milk. It was shown that whey proteins form a continuous and coherent film at the surface of cotton fibers. The authors demonstrated that the whey protein coating induced an enhancement of cotton charring during degradation under nitrogen atmosphere. Proteins retain a high water amount that provokes hydrolysis and thus amino acids are released. Carboxylic groups catalyze the cellulose decomposition favouring dehydration and thus charring to the detriment of depolymerization. With regard to fire properties, it was evidenced that protein coating enables to slow down the burning rate and hence to increase the burning time. The authors investigated also the influence of protein denaturation. It was noted that the effect on thermal and fire properties was the same whether the macromolecules were folded or unfolded. Alongi et al. [42] carried out a similar study using this time caseins and hydrophobins as coating deposited on cotton fabrics. Caseins are the main proteins of milk and belong to the phosphoprotein group meaning that some amino acid residues bear phosphate moieties as it was presented in Fig. 2.5. Hydrophobins are proteins produced by filamentous fungi, they exhibit an amphiphilic character and molar masses. Whatever the kind of proteins, the effects were similar to those obtained with whey proteins with catalysis of cellulose thermal decomposition and an increase of char residue at high temperature. Regarding to fire properties, the early degradation of cellulose induces a drastic decrease of time-to-ignition. On the other hand the formation of char layer enables a decrease of the pHRR and a slowdown of the burning rate.

In a similar way, Alongi et al. [43] used DNA from herring sperm as flame retardant and suppressant for cotton fabrics. The authors claimed that DNA was a natural intumescent flame retardant since the three main conditions for intumescence are met: a sugar unit acting as carbon source, a phosphate group able to release phosphoric acid and nitrogen-containing bases likely to release ammonia that plays the role of blowing agent [44]. After treatment, DNA forms a homogeneous and continuous coating at the cellulose fiber surface. This coating tends to accelerate the thermal decomposition of cellulose that starts from 200 °C. This early degradation was attributed to the release of phosphoric acid by DNA. Phosphoric acid catalyzes dehydration of cellulose and promotes the formation of an aromatic char thus limiting the production of fuel such as levoglucosan [45]. The authors assessed the flammability of the DNA-treated cotton fabrics using a horizontal burning test and cone calorimeter. Above a 10 wt% add-on, the DNA coating enabled the extinguishment of cotton as soon as the methane flame was removed. In cone calorimeter a 19 wt% add-on prevents the ignition to occur [46]. In this latter case, it was proved that the DNA coating acts as a protective barrier that absorbs heat and releases inert gas. Thus the underlying cellulose undergoes pyrolysis instead of burning. It was also evidenced that the fire performance of coatings depends on the DNA molecular size and pH of the impregnation solution and number of impregnations [47]. Improving the fire behavior of cotton fabrics may also be achieved by using DNA in Layer by Layer (LbL) assemblies. Carosio et al. [48] proposed a bilayer composed of DNA as polyanion and chitosan as polycation. 5–20 bilayers were deposited onto the cotton fabrics. DNA layers promote the charring of chitosan. The obtained char is thermally stable and imparts a remarkable flame retardant character to cotton.

Carosio et al. [49] used DNA LbL assembly to improve the fire behavior of a polyethylene terephthalate (PET) foam. In this work, a DNA aqueous solution was prepared and was applied as one of the anionic layers in the LbL assembly, Poly (diallydimethylammonium chloride) being the cationic layer and poly(acrylic acid) branched polyethylene imine being the other anionic layer. A LbL assembly containing ammonium polyphosphate (APP) instead of DNA was prepared for comparison. Flammability and cone calorimetry tests revealed that the LbL coatings containing APP induces superior performances. These results were ascribed to the fact that only APP-based architectures were able to suppress the melt dripping behavior typical of PET and hence to reduce the heat release rate peak.

DNA can also be utilized for the flame retardancy of thick polymers. For this application two strategies have been investigated. The first one consists in incorporating DNA in the bulk by melt-blending, in the second one DNA is deposited as coating onto the thick polymeric sample. When blended with EVA, DNA was proved to change the degradation pathway of the polymer leading to the formation of char, inhibiting the production of volatiles [38]. At a 20 wt% content DNA induces a 39% decrease of the pHRR (Fig. 2.11) and a strong reduction of CO and CO_2 production. It was concluded that DNA acts similarly to APP. When confined at EVA surface, the DNA coating was highlighted to block ignition when tested with cone calorimeter under a 35 kW/m^2 heat flux [50]. Hence ignition was greatly delayed and the combustion kinetics was drastically reduced. The coated EVA can also withstand the exposure to a butane/propane flame. Good results were obtained when DNA coating was applied onto other thermoplastic polymers (PP, polyamide 6 (PA6), ABS, polyethylene terephthalate (PET)) [51].

2.3.2 Bio-resources Combined with Phosphorus or Nitrogen Compounds

Considering the intrinsic charring ability of some bio-resources as described in paragraph 2, a very frequent strategy for developing performant flame retardant systems consisted in combining biomolecules with traditional flame retardant known for their propensity to emphasize charring or to expand carbonaceous residues. That is probably why most of the studies presented hereafter deals with combination of renewable resources with phosphorus or nitrogen containing compounds.

2.3.2.1 Carbohydrates

Carbohydrates are oxygen containing polymers and therefore may be used as charring agent in flame retardant compositions. Among carbohydrates starch was more particularly studied as carbon source in intumescent systems. In 2008 Réti et al. [52] used potato starch as substituent to pentaerythritol (PER) in intumescent composition. 10 wt% starch was combined to 30 wt% ammonium polyphosphate (APP) in a polylactic acid (PLA) biopolymer. This enables to obtain a limiting oxygen index (LOI) of 40% and a V-0 rating at the UL94 test. This later rating was even better than that obtained with the PER containing composition (V-2). With regard to cone calorimeter test, the starch containing formulation enables to increase the time-to-ignition. The peak of heat release rate was also decreased but at lesser degree compared to the APP/PER system. The effect of starch was related to the high amount of charred residue (circa 50 wt%) generated by its interaction with APP. This char acts as a protective layer for the underlying polymer during the combustion.

Wang et al. [53] studied a similar starch based flame retardant system in PLA but this time APP was microencapsulated in polyurethane and associated with melamine. They observed the same trends than Reti et al. i.e. an increase of LOI, a V-0 rating in UL94 test and a decrease of pHRR measured with microscale combustion calorimeter (PCFC). It was evidenced that the FR system composed of microencapsulated APP melamine and starch has an excellent intumescent effect. The authors claimed that microencapsulation offers two main advantages: (i) an improved compatibility of APP with the polymer matrix, (ii) a retardation of the reaction between acid and carbon source during processing.

Another strategy to take benefit from starch in the improvement of fire properties was to use it as part of the polymer matrix. Gaialene is a polyolefin grafted starch developed by Roquette at the beginning of the 2010's with the aim to propose a lower carbon footprint polymer. Dupretz et al. [54, 55] investigated the fire behavior of Gaialene when associated with APP or APP and melamine. When used alone, APP improves very significantly the fire performance of Gaialene reaching a LOI of 30%, a V-0 rating at UL94 test and a 65% reduction of pHRR. The dehydrating effect of phosphoric acid released from APP provokes depolymerization and aromatization of starch leading to charring. The release of different gaseous products and the decrease of the viscosity generate the swelling of the structure giving rise to an efficient barrier effect. That makes it a good candidate for replacing PP/APP/PER intumescent formulation.

Starch can be also employed in flame retardant coating for textiles using the so-called layer by layer (LbL) technique. Carosio et al. [56] used a starch solution as cationic electrolyte as well as carbon source and a solution containing poly (phosphoric acid) (PPA) was anionic electrolyte and acid source. 2 or 4 bilayers were deposited onto cotton fabrics of different densities. Despite the limited number of bilayers this FR system was likely to generate a strong charring that enables a self-extinguishing behavior. When exposed to the radiant heat of the cone calorimeter the total heat released by the FR fabrics was considerably reduced compared to the uncoated one.

Chitosan which is the deacetylated form of chitin, is another compound belonging to carbohydrates. Chitosan has raised a great interest in FR formulation during the last five years. The major part of papers dealing with chitosan is dedicated to the layer by layer (LbL) deposition technique. In acidic solution chitosan becomes a polycation, thus it may play the role of the positively charged layer. Chitosan was associated with various negatively charged layers. Carosio et al. [57] carried out the coating of PET-cotton fabrics using chitosan/APP bilayers that they compared with silica/APP bilayers. The LbL architectures was shown to enhance the overall flame retardancy of the fabric blend. The most significant result concerned the suppression of the afterglow phenomenon. This was associated with a significant increase of the residue after the test. Pursuing this work, Alongi et al. [58] used chitosan in bilayers (chitosan/APP) and quadlayers (chitosan/APP/silica/ silica). Combination of bilayers was almost equivalent to quadlayers. The development of a coherent and homogeneous coating enables to slow down the thermal decomposition of the fabrics. The flame retardant mechanisms of APP/chitosan LbL assembly was studied in details by Jimenez et al. [59] when deposited on cotton fabrics. The LbL assembly promotes the formation of a protective char layer by dehydration of both chitosan and cotton. In the gas phase, the release of water tends to extinguish the flame while other volatiles favor a kind of "microintumescence" phenomenon. Hu's team in Hefei developed various LbL assemblies involving chitosan to fireproof cotton fabrics. Hence chitosan was successively associated with phosphorylated cellulose (P-Cell) [60], phosphorylated chitin (P-Chit) [61], phosphorylated polyvinyl alcohol (P-PVA) [62], titanate nanotubes (TiNT) [63]. Generally speaking LbL assemblies tend to decrease the thermal stability of the fabrics except with titanate nanotubes (Fig. 2.13) but in return promote charring. The char yield increases with increasing bilayers and depends on the respective concentrations of the layers. Chitosan based coatings decrease the burning time of fabrics and suppress afterglow. pHRR and THR are significantly decreased (Fig. 2.14).

Chitosan based LbL assemblies were also used for the flame retardancy of polyurethane foams. Carosio et al. [64] developed a LbL chitosan/poly(allylamine diphosphonate) coating that was applied on an open cell PU foam. The coating enables preventing the melt dripping phenomenon while burning. It can also drastically reduce the heat release rate in cone calorimeter test. The assembly can efficiently hinder the foam structural collapsing during combustion. 2 bilayers were sufficient to reduce the pHRR by 48%. Pan et al. [65] realized by LbL method a fully biobased coating containing chitosan and lignosulfonate on the surface of flexible polyurethane foam. The main result concerns the reduction by 42% of the

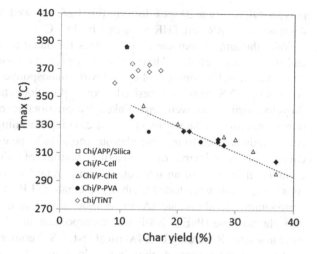

Fig. 2.13 Temperature of the pHRR in PCFC versus char yield for cotton flame retarded by various LbL assemblies based on chitosan from [58, 60–63]

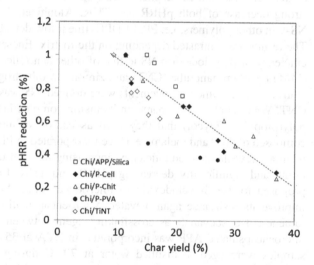

Fig. 2.14 Relative variation of pHRR versus char yield for cotton flame retarded by various LbL assemblies based on chitosan from [58, 60–63]

pHRR with 8 bilayers. Zhang et al. [66] used chitosan in combination with phytic acid in a polyelectrolyte complex (PEC) for the flame retardancy of ethylene vinyl acetate (EVA). Polyelectrolyte complexes differ from LbL assemblies in the sense that oppositely charged compounds are combined in a unique solution. After drying the PEC was incorporated in EVA at loading varying from 0 to 20 wt%. The presence of PEC enables increasing the char yield at high temperature. Furthermore, introduction of PEC leads to a reduction of the pHRR and THR measured by PCFC. Highly expanded and stable charred foam was formed generating an obvious barrier effect during combustion. By the same token Zhang et al. [67] prepared a PEC based on polyethylene imine and phytic acid for the flame retardancy of polypropylene. At a 20 wt% loading this PEC allowed for a LOI of 25.1% and a decrease of pHRR and THR measured by PCFC.

With the aim to combine two strategies for flame retardancy, i.e. intumescence and use of nanoparticles, Alongi et al. [68] prepared a new flame retardant system containing a stable complex of cyclodextrin nanosponges (NS) and a phosphorous compound. NS was synthesized from β-cyclodextrin a starch derivative. β-cyclodextrin units were crosslinked by an organic carbonate thus forming a porous structure including two types of cavities. Grinding of NS and phosphorous compounds enabled to the phosphorous moieties to be entrapped either in internal cavities of cyclodextrin or in external cavities of NS. The resulting complex exhibits all features of an intumescent flame retardant system. Different phosphorous compounds were tested triethyl phosphate (TEP), triphenyl phosphate (TPP), ammonium polyphosphate (APP), dibasic ammonium phosphate (APb), diethyl phosphoramidate (PhEt). NS-P was incorporated in EVA and the fire properties were assessed. Regarding UL94 vertical test, NS-P confers a V-2 rating whatever its content and the source of phosphorus. In cone calorimeter test, NS-P enables a strong decrease of both pHRR and THR. Alongi et al. evaluated the interest of NS-P in other polymers, i.e. PP, LLDPE (linear low density polyethylene) and PA6. The results are contrasted depending on the matrix. Enescu et al. [69] compared the efficiency of β-cyclodextrin NS to that of other nanofillers such as montmorillonite (MMT), carbon nanotube (CNT) and zirconium dihydrogen phosphate in a PA6,6 matrix. Even if some improvements were noticed, NS was supplanted by MMT and CNT. Wang et al. [70] prepared an inclusion complex between β-cyclodextrin and poly(propylene glycol) that they used as carbon in intumescent flame retardant composed of APP and melamine. Once incorporated in PLA, this intumescent flame retardant revealed a good efficiency, increasing the LOI up to 34%, conferring V-0 rating and significantly decreasing pHRR and THR. Later on Wang et al. [71] proposed to use β-cyclodextrin to microencapsulate APP with the final goal to improve its resistance against water. Microencapsulation was achieved by using toluene-2,4 diisocyanate as crosslinking agent between β-cyclodextrin units. The microencapsulated APP was incorporated in EVA at 35 and 40 wt% loading. The samples were aged in distilled water at 70 °C during various periods of time. Cyclodextrin capsules were highlighted to increase the durability of APP since V-0 rating was retained after 96 h of ageing while the composite with non-protected APP was non-rated after 24 h of ageing.

Saponins are a class of biomolecules found in various plants that belong to amphiphilic glycosides. They arise from the condensation of a sugar and another functional group. In saponins sugars are bonded to cyclic triterpene. Recently tea saponin extracted from the nutshell of camelia was used as both blowing agent and carbon source in an intumescent flame retardant formulation [72]. Tea saponin was mixed with APP and pentaerythritol (PER) to give an intumescent flame retardant (IFR) that was thereafter blended with an alkyd varnish and deposited onto wood boards. During combustion an intumescent layer was formed that enabled a decrease of pHRR and THR measured by cone calorimeter. In the same manner tea saponin was associated to APP and melamine to form a new intumescent flame retardant system for natural rubber [73]. This system was of interest because it enabled to combine both good fire properties and reinforced mechanical properties.

2.3.2.2 Phenolic Compounds

It was shown in a preceding section that lignin by itself can bring flame retardant properties to polymers. However some weak points e.g. low thermal stability, poor cohesion of char residue can be pointed out. That is why lignin has often been used in synergy with traditional flame retardant. De Chirico et al. [34] used lignin in combination with aluminum hydroxide (ATH), melamine phosphate, mono-ammonium phosphate and APP in a polypropylene (PP) matrix. These combinations led generally to increase the thermal degradation temperature, the combustion time and the char yield of PP, and decrease the rate of heat release and the weight loss rate during combustion. Réti et al. used lignin as substituent for pentaerythritol in APP based intumescent composition. PLA loaded with 30 wt% APP and 10 wt% lignin reached a V-0 rating at UL94 test instead of V-2 for the PER containing formulation. Nevertheless the APP/PER system remained more performant in cone calorimeter and LOI tests. More recently the same flame retardant system (APP/lignin) was utilized by Cayla et al. [74] for fireproofing PLA fabrics. An optimization of formulation was performed in order to obtain both good spinnability and satisfying fire properties. It was evidenced that 5 wt% of lignin and 5 wt% of APP were sufficient to obtain an efficient FR effect in PLA fabric.

Tannins are other polyphenolic compounds that may be of relevance for flame retardancy. However this is so far an almost unexplored area. Tondi et al. [75] used mimosa tannin extract combined with hexamine as hardener for the impregnation of wood. Boric acid and phosphoric acid were added as synergist in the tannin based solution. Fire behavior was assessed by exposure to the flame of Bunsen burner. It was highlighted that ignition of wood was strongly delayed after treatment. Very close results were obtained whatever the synergist used with tannin.

2.3.3 Modified Bio-resources with Enhanced Charring Effect

The preceding paragraph has shown that natural raw matter may advantageously be combined with phosphorus or nitrogen compounds to enhance charring. Following the same objective, many studies have been devoted to the chemical modification of bio-resources with the aim to build new all-in-one biobased flame retardant molecules. Among the possible chemical modification, phosphorylation was probably the most frequently investigated. Phosphorylation of biobased compounds was the subject of a review by Illy et al. [76] that described very completely all the synthetic routes than can be used. Nevertheless other types of chemical modification may also be encountered.

2.3.3.1 Carbohydrates

Numerous works deal with phosphorylation of cellulose within the framework of textiles flame retardancy. However cellulose has rarely been modified with the specific aim to build a biobased flame retardant. In 2010 Aoki and Nishio [77] modified a cellulose ester derivative with various phosphoryl groups. Depending on the moiety, they observed a significant increase of the char yield at 700 °C under nitrogen atmosphere up to 29 wt% for diethylphosphoryl group. The phosphorylated cellulose derivatives were incorporated in polylactic acid (PLA) and polyethylene terephthalate (PET) and a V-2 ranking was obtained at the UL94 vertical test. Pan et al. [78] prepared a cellulose derivative containing both phosphorus and nitrogen. Firstly the microcrystalline cellulose (MCC) was phosphorylated by phosphorous acid in molten urea and then the phosphorylated cellulose was reacted with ammonia water to obtain the final product (CPA). The biobased flame retardant was incorporated into polyvinyl alcohol (PVA) at contents varying from 0 to 15 wt%. The presence of CPA lowered the thermal stability of composites and the early emission of phosphoric acid favored dehydration of cellulose at the expense of depolymerization. Thus, the char yield was enhanced up to 19.4 wt% for the highest loading. Microcalorimetry of combustion revealed that pHRR as well as THR were reduced. At 15 wt% loading PCA enabled PVA to obtain V-0 rating at UL94 test and LOI of 30%. More recently Costes et al. [79] prepared a phosphorylated MCC using the same procedure than Pan et al. This modified cellulose was introduced at 20 wt% loading for PLA fireproofing. Despite a V-0 ranking at vertical burning test, the cone calorimeter quantities (pHRR and THR) were only slightly reduced. Thereafter the authors decided to substitute a part of phosphorylated MCC by aluminum phytate a derivative of the biobased phytic acid. This time the results were much more convincing. It was assigned to the charring effect of aluminum phytate that leads to the fast formation of the charred layer. Phosphorylated cellulose can also be combined with other biobased compounds via the layer by layer (LbL) process. Pan et al. [60] proposed a LbL assembly composed of chitosan as polycation and phosphorylated cellulose as polyanion. The ionic character was given by adjusting the pH of the solutions. This LbL assembly was used for the fireproofing of cotton fabrics. It was noticed that the char yield depended on the proportion of the two components and the number of bilayers. When coated with 20 bilayers the cotton fabrics was self-extiguishing and the pHRR and THR measured by PCFC were drastically reduced. A similar study was done by substituting phosphorylated cellulose by phosphorylated chitin (another carbohydrate) [61]. In both cases, the high fire performance was attributed to the insulating properties of the phosphorus containing char formed at the cotton surface. Among the various ways to achieve cellulose phosphorylation, one is of particular interest within the frame of green chemistry. Božič et al. [80] studied the enzyme-mediated phosphorylation of cellulose nanofibers (CNF) by using hexok-inase and adenosine-50-triphosphate in the presence of Mg-ions. Depending on the operating conditions, up to 8.8 wt% of phosphorus may be added to cellulose. In this latter case the char residue of phosphorylated CNF is circa 57 wt% at 600 °C which is remarkable. This offers potential applications for cellulose modified by enzymatic means to be used as flame retardant.

Hu and his coworkers carried out several studies involving modified chitosan. The first step of all modifications consisted in the esterification of chitosan by phosphorous pentoxide (P_2O_5). In a second step different modifications of the phosphorylated chitosan (PCS) were envisaged. In a first paper PCS was reacted with nickel nitrate hexahydrate to give a nickel chitosan phosphate (NiPCS) [83]. In a second article a urea salt of chitosan phosphate was prepared (UPCS) [82]. In a third article a melamine salt of chitosan phosphate was synthesized (MPCS) [81]. Finally in a last paper, PCS was reacted with glycidyl methacrylate to obtain a chitosan phosphate acrylate (GPCS) [84]. In the first three studies the chitosan based flame retardants were incorporated in PVA while in the last case, the chitosan phosphate methacrylate was used in an epoxy acrylate resin. Whatever the system, the conclusions were roughly the same. The presence of the phosphorous moieties accelerates the dehydration of chitosan and therefore promotes charring (Fig. 2.15). The char yield were significantly increased while the quantities (pHRR and THR) measured by micro-calorimetry of combustion were decreased (Fig. 2.16).

Howell et al. [85, 86] investigated the potential of low cost by-products from the wine industry as raw materials for flame retardant. Hence tartaric acid was converted into its diethyl ester form. The phosphorylation of diethyl tartrate was then achieved using diphenyl phosphinic chloride. The resulting diethyl 2,3-diphenylphosphinato-1,4-butanedioate may be used as flame retardant since its thermal decomposition is likely to liberate diphenylphosphinic acid to promote char formation, at the combustion temperature of several common polymers.

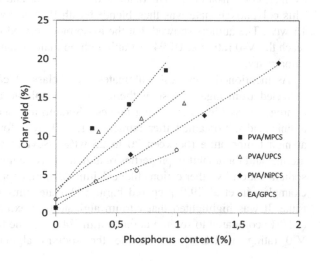

Fig. 2.15 Char yield versus phosphorus content for PVA and EA flame retarded by various phosphorylated chitosan from [81–84]

Fig. 2.16 Relative variation of pHRR versus char yield for PVA flame retarded by various phosphorylated chitosan from [81–84]

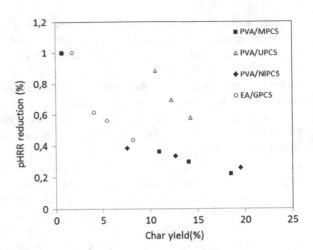

Howell and Daniel [87] also investigated the thermal degradation of a series of phosphorus esters derived from a diol generated by the esterification of isosorbide (from starch) with 10-undecenoic acid (from castor oil) followed by thiol-ene reaction with 2-hydroxyethanethiol. The diol could thereafter be converted to phosphorus esters either by treatment with phosphoryl chloride in the presence of an acid acceptor or by treatment with a phosphite in carbon tetrachloride. It was evidenced that the thermal stability of the esters was extremely dependent on their chain ends. The diethyl phosphate-terminated ester exhibits a very low degradation onset while the DOPO-terminated ester was very stable (DOPO: 9,10-dihydro-9-oxa-10-phosphahenanthrene-10-oxide). The char yield was also dependent on the structure and with regard to this parameter the diphenyl phosphate terminated-ester was the most performant. Isosorbide was also used by Mauldin et al. [88] to prepare biobased flame retardant system. In this work, a polyphosphonate was synthesized by melt condensation polymerization of isosorbide and a phosphonic dichloride. This polyphosphonate was then blended with PLA at loadings varying from 0 to 15 wt%. The authors showed that the isosorbide derived flame retardant enables to reach the V-0 rating at UL94 test although no major change was observed by cone calorimetry.

As mentioned previously, alginates are a class of carbohydrates that can be extracted from algae. Most of them are water soluble and thus can be used as hydrogel or as film after casting process. Sodium alginate is the common form of alginic acid salt obtained after extraction. It is likely to form a large amount of char at high temperature (between 20 to 25 wt%). Several works evidenced that the thermal and flammability properties of alginate may be improved by exchanging the sodium cation by other cations like calcium, barium, copper or zinc [89–91]. As an example Liu et al. [92] prepared barium alginate films by a facile ion exchange route. It was highlighted that barium alginate film exhibits very high LOI value (52.0%) compared to sodium alginate film (24.5%). The biobased film obtained the V-0 rating at UL-94 test, while the sodium alginate film was not rated.

Moreover PCFC tests revealed a strong decrease of pHRR for barium alginate film. It is suggested that the barium ion considerably modified the degradation pathway of alginate by catalytic effect that promotes the formation of more stable products.

Phytic acid or phytates are the phosphorous containing biomolecules naturally found in plant tissues. Sodium phytate is the most common commercial form of phytates. Costes et al. [93] investigated the flame retardant effect of various metallic phytates in PLA. Following an approach close to that above-described for alginates the authors have substituted the sodium cation by aluminum, iron and lanthanum cations using a solubilization/precipitation procedure. Metallic phytates were incorporated in PLA at loading varying from 0 to 30 wt%. Blends of phytates were also tested. Phytates were shown to promote the rapid formation of a charred layer during combustion. This layer induced a significant reduction of pHRR. The best results were obtained with aluminum phytate. Unfortunately this compound was proved to be responsible for the degradation of PLA during melt processing. A solution to this issue was found by combining two phytates.

2.3.3.2   Phenolic Compounds

Lignin was highlighted to be interesting raw material for biobased flame retardant since it is a low cost and intrinsically high charring product. Hence many attempts were made to modify polyaromatic biomolecule with the aim to enhance its fire-proofing efficiency. In 2012 Yu et al. [94] performed the grafting of a phosphorus-nitrogen compound onto lignin. The first step of the modification consists in the hydroxymethylation of lignin using formaldehyde. The hydroxymethylated lignin was then reacted with a phosphoryl dichloride imidazole to obtain PN-lignin. This modified lignin was incorporated into polypropylene at loading up to 30 wt%. An increase of both the thermal stability and the char yield was proved by thermogravimetric analysis. In the cone calorimeter test, the increase of the residue was associated to a decrease of pHRR and THR. The better fire performance of PN-lignin compared to unmodified lignin was assigned to the quality of the charred layer that was shown to be more continuous and compact. Ferry et al. [36] studied the grafting of different molecular or macromolecular phosphorous compounds onto kraft lignin. Thus dihydrogen ammonium phosphate, a homopolymer of (methacryloyloxy) methyl phosphonic acid and a copolymer of (methacryloyloxy) methyl phosphonic acid and methyl methacrylate were used as grafting agents. The grafted lignin was blended with polybutylene succinate (PBS) at content varying from 0 to 20 wt%. The modification of lignin did not impart enhancement of char yield. Nevertheless slight improvement in cone calorimeter quantities (pHRR and MAHRE) was observed. As did Yu et al., this was attributed to the higher quality of the char layer which was more homogeneous and cohesive. A simple route to achieve the phosphorylation of lignin was proposed by Prieur et al. [95]. Lignin was reacted with phosphorous pentoxide (P_2O_5) in tetrahydrofuran (THF). The phosphorylated lignin (P-Lig) was incorporated in Acrylonitrile Butadiene Styrene (ABS) at 30 wt% loading. The phosphorus content in P-Lig was measured to be 4.1 wt%. P-Lig induced a huge increase of char yield which was as high as 55 wt%. In cone calorimeter test, despite a decrease of time-to-ignition, the results were satisfying with a decrease of pHRR and THR stronger than with pure lignin. More recently Costes et al. [37] studied a two-step phosphorus/nitrogen chemical modification of lignin to enhance the fire behavior of PLA. This approach was applied to both kraft and organosolv lignins. The modified lignin was blended with PLA at a 20 wt% loading. Thermal analysis revealed that modified lignin increases the temperature of the maximum mass loss rate. However the char yield was lower than that obtained with unmodified lignin. The modified lignins were proved to be highly effective to reduce the flammability of PLA composites and V-0 rating was reached at UL-94 test. The most interesting result was probably the fact that phosphorus-nitrogen grafting enables to counterbalance the thermal destabilization generally induced by lignin and to maintain a high time-to-ignition. Liu et al. [96, 97] performed the modification of lignin using polyethyleneimine (PEI) diethyl phosphite and metal acetate. This enables binding phosphorus and nitrogen atoms as well as metallic cations (Zn^{2+} or Cu^{2+}) to lignin. The functionalized lignin was used either in PBS at loading varying from 0 to 10 wt % or in PP/wood composite at content up to 15 wt%. The results were much probative in PBS with a large increase of char yield and decrease of pHRR and THR. Metal cations were supposed to catalyze the carbonization of polymer and lignin leading to higher residue and better flame retardancy.

Phosphorylation is not the only pathway for confering to lignin enhanced flame retardant properties. Zhang et al. [98] prepared a urea-modified lignin (UM-Lig) according to the Mannich reaction. This nitrogen containing lignin was combined with ammonium polyphosphate (APP) and used as a novel intumescent flame-retardant for PLA. Depending on the proportion between APP and UM-Lig it was possible to reach V-0 rating in vertical burning test and a LOI of 34.5%. pHRR and THR were drastically reduced due to the formation of a thick intumescent char that represents up to 35% of the initial sample mass. A second strategy tested by Zhang et al. consists in preparing a lignin-silica hybrid (LSH) by the sol–gel method [99]. Combined with APP, LSH gives a very performant intumescent flame retardant system for PLA that leads to effects very similar to those obtained with UM-Lig.

Apart from lignin other natural phenolic compounds may be used to prepared flame retardant additives. In 2002 Marosi et al. [100] demonstrated the possibility to achieve phosphorylation or phosphinylation of phloroglucinol and hydroquinone, two biobased hydroxyphenols. Phosphorylation involved the reaction with chlorophosphates while phosphinylation implies phosphinyl chloride. Later Vothi et al. [101] achieved the phosphorylation of phloroglucinol and resorcinol by various chlorophosphates. The obtained cyclic phosphates were introduced at 30 wt % loading in ABS and 5 wt% loading in polycarbonate (PC). The phosphorous flame retardant promotes polymer charring. The best results were obtained with the phloroglucinol triphenyl phosphate with a V-0 rating when used at only 3 wt% in PC. More recently Ménard et al. [102] synthesized various phloroglucinol phosphates according to a protocol similar to that of Vothi et al. These biobased flame

Fig. 2.17 Char yield versus phosphorus content for epoxy resins flame retarded by additive or reactive flame retardant based phosphorus modified phloroglucinol from [102–104], ■ DGEBA-IPDA-P3P(OEt), △ DGEBA-IPDA-P2EP1P, ● DGEBA-IPDA-P3SP, ◇ DGEBA-IPDA –P2EP1SP, □ P3EP-IPDA-P2EP1P, ▲ P3EP-DA10-P2EP1P, ○ P3EP-DIFFA-P2EP1P

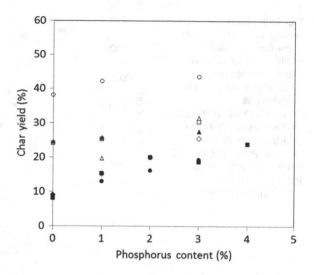

retardants were introduced in an epoxy resin (bisphenol A diglycidyl ether/isophorone diamine hereinafter DGEBA/IPDA) and the global phosphorus content was varied between 0 and 3 wt%. Whatever the molecule the char yield was shown to increase with increasing phosphorus content (Fig. 2.17).

In cone calorimeter test a spectacular decrease of pHRR and THR was observed with the phloroglucinol diethyl phosphate (P3P(OEt)). This was associated with a huge expansion of the char residue during the combustion forming a highly thermally insulating layer at the sample surface. Ménard et al. [104] prepared also phosphonate flame retardant from phloroglucinol. The synthesis consists firstly in the glycidylation of phloroglucinol and secondly in the reaction of epoxydized phloroglucinol with diethyl(3-mercaptopropyl) phosphonate. The interest of the study lies in the comparison of additive (P3SP) and reactive (P2EP1SP) flame retardants with similar structure. It was concluded that both approaches are very efficient in reducing pHRR and THR (Fig. 2.18). The reactive flame retardant was shown to enable a higher charring effect and a higher intumescence.

2.3.4 Reactive Biobased Flame Retardants

There are various motivations to use reactive rather than additive flame retardants. In their review about the effects of ageing on the fire behavior of flame retarded polymers Vahabi et al. [105] mentioned several situations where ageing induces migration of the flame retardant additive out of the polymer matrix onto the sample surface. The so-called blooming or exudation mechanism may be due to the low solubility and/or the small size of the additive and results generally in a decrease of the fire performance. Therefore, a way to counteract this effect consists in inserting

Fig. 2.18 pHRR (PCFC) versus char yield for epoxy resins flame retarded by additive or reactive flame retardant based phosphorus modified phloroglucinol from [102–104], ■ DGEBA-IPDA-P3P(OEt), △ DGEBA-IPDA-P2EP1P, ● DGEBA-IPDA-P3SP, ◇ DGEBA-IPDA –P2EP1SP, □ P3EP-IPDA-P2EP1P, ▲ P3EP-DA10-P2EP1P, ○ P3EP-DIFFA-P2EP1P

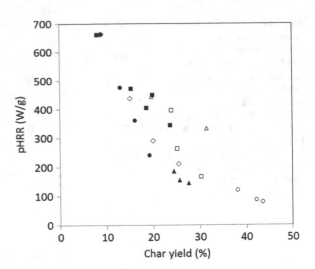

covalently the flame retardant into the polymer chain. Another reason for using reactive flame retardants may be to improve their efficiency by ensuring a good dispersion at the molecular scale. Bio-oils and bio-phenols were the two resources used to prepare biobased reactive flame retardants.

2.3.4.1   Oil Derivatives

The group of Lligadas in Spain carried out several studies where they tried to prepare functional polymers from vegetable oils [106, 107]. They succeeded in synthesizing a biobased silicon-containing polyurethane, with a silicon content between 1.7 and 9.0 wt%, with interesting thermal properties [107]. The starting raw matter is an ester of ω-unsaturated fatty acid easily obtained from natural ricinoleic acid. In a first step the methyl 10-undecenoate was hydrosilylated by phenyltris(dimethylsiloxy) silane and then reduced to obtain a silicon containing polyol. In a second step this polyol was reacted with a di-isocyanate to obtain an elastomeric polyurethane. By TGA it was shown that the mass loss was lower in the range 300–600 °C in the presence of silicon even if the high temperature residue was almost constant whatever the silicon content. The limiting oxygen index (LOI) was highlighted to increase from 18.2 to 23.6% when the silicon content increases from 0 to 9 wt%. It was demonstrated that a continuous layer of silica was formed at the sample surface retarding the oxidation of the char. Using the same bioresource, Lligadas et al. prepared two epoxy monomers: an epoxidized 10-undecenoyl triglyceride (Fig. 2.19) and an epoxidized methyl 3,4,5-tris(10-undecenoyloxy) benzoate. 4,4′-diaminodiphenylmethane (DDM) or bis(m-aminophenyl)methylphosphine oxide (BAMPO) were used as crosslinking agent. The flame retardancy of these epoxy resins was achieved by the use of a phosphorus containing epoxy monomer. This

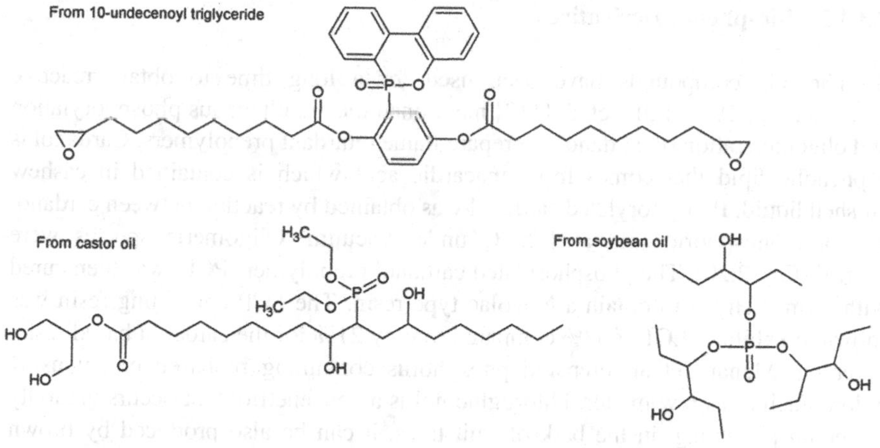

Fig. 2.19 Oil-based reactive flame retardants

monomer was synthesized from the ω-unsaturated fatty acid derivative and from a DOPO derivative [108]. The presence of phosphorus moieties enabled to increase the char residue (at 800 °C) up to 18 wt% for 5.7 wt% phosphorus content. Flammability was significantly improved and a tight correlation was evidenced between LOI and the phosphorus content into the epoxy resin [109].

Heinen et al. [110] developed oil-based rigid polyurethanes with enhanced fire properties. In order to obtain phosphorylated polyols (Fig. 2.19), an epoxidized soybean oil was reacted with phosphoric acid. Polyurethane foams were then prepared by reacting the above-mentioned polyols (after neutralization with triethanolamine) with polymeric diphenylmethane diisocyanate (MDI) while pentane was used as blowing agent. Light foams with homogeneous cell size distribution were obtained. LOI was shown to increase with increasing phosphorus content in the foam. The best performance measured with reactive biobased monomer was comparable to that obtained with a commercial phosphorous flame retardant. It was evidenced that the fireproofing action took place mainly in the condensed phase by increasing the char residue. An approach similar to that of Heinen et al. was used by Zhang et al. [111] to obtain flame retarded polyurethane foams. This time castor oil was used as raw matter. It was alcoholized with glycerol (GCO) and then epoxidized to obtain a polyol that was further reacted with diethyl phosphate (see Fig. 2.19). The resulting phosphorylated polyol (COFPL) was mixed with GCO and reacted to methylene diphenyl 4,4′-diisocyanate to form a rigid polyurethane foam containing various amount of phosphorus. It was highlighted that LOI increases with increasing phosphorus content up to 24.5 wt% when COPFL was used as unique polyol. The cone calorimeter test revealed that the phosphorylated polyol enabled a slight decrease of pHRR and more particularly a significant decrease of THR that was attributed to the promotion of a thick and compact char.

2.3.4.2 Bio-phenol Derivatives

Bio-phenolic compounds have been used for a long time to obtain reactive monomers. In 1990, Pillai et al. [112] mentioned the simultaneous phosphorylation and oligomerization of cardanol to prepare flame retardant prepolymers. Cardanol is a phenolic lipid that comes from anacardic acid which is contained in cashew nutshell liquid. Phosphorylated cardanol was obtained by reaction between cardanol and orthophosphoric acid at 175 °C under vacuum. Oligomeric species were formed (Fig. 2.20). The phosphorylated cardanol prepolymer (PCP) was then cured with formaldehyde to obtain a Novolac type resin. The PCP containing resin was shown to exhibit a LOI of 33% compared to only 21% for the cardanol based resin. Later on, Menard et al. prepared phosphorus containing biobased epoxy using phloroglucinol as raw matter. Phloroglucinol is a benzenetriol that occurs naturally in certain plants e.g. in the bark of fruit trees, it can be also produced by brown algae or bacteria. In a first step epoxydization of phloroglucinol was achieved by reaction with epichlorhydrin. It leads to the triglycidyl phloroglucinol (P3EP) that can be further used as epoxy monomer. In a second step P3EP was reacted with triethyl phosphite with zinc chloride as catalyst to obtain the diglycidyl mono-phosphonated phloroglucinol P2EP1P (see Fig. 2.20) which can be considered as a reactive phosphorous flame retardant [103]. Another method consists in reacting P3EP with the diethyl(3-mercaptopropyl)phosphonate to obtain the diglycidyl thiophosphonated phloroglucinol which is a sulfur and phosphorus containing reactive flame retardant [104]. The two reactive monomers were used for fireproofing a DGEBA/IPDA epoxy resin. In both cases it was shown that the use of reactive flame retardants induces a strong increase of char residue that was almost proportional to the phosphorus content in the resin (Fig. 2.17). Moreover, a huge

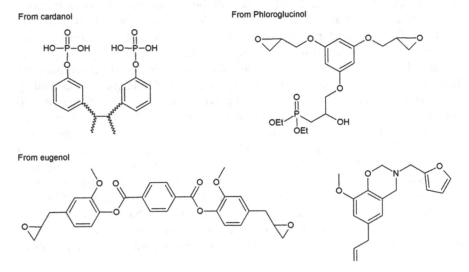

Fig. 2.20 Biophenol-based reactive flame retardants

intumescent effect was highlighted. This results not only in a decrease of pHRR (Fig. 2.18) but also in a reduction of THR. By combining P3EP as epoxy monomer, P2EP1P as reactive flame retardant and difurfurylamine a commercial biobased diamine the authors succeeded in preparing a fully biobased flame retarded epoxy resin.

Eugenol is another biophenol naturally found in cloves that has been used as raw matter for developing reactive flame retardant. Eugenol was reacted with terephthaloyl chloride to obtain a diene intermediate [113]. Thereafter this diene was epoxidized by reaction with a peracid. Thus an aromatic-ester type eugenol based epoxy monomer was obtained (Fig. 2.20). This monomer was formulated in a DGEBA-DDS resin. It was highlighted that the eugenol based monomer enables an increase of LOI from 23.5 to 26.8. Moreover, it provokes the rapid extinguishment of flame in a vertical burning test. In cone calorimeter test, despite a lower time-to-ignition, the eugenol monomer allows a decrease of pHRR as well as THR. This set of performance was ascribed to the ability of the aromatic monomer to promote a char layer that acts as protective shield during fire tests. Eugenol was also used by Thirukumaran et al. [114] as raw material to synthesize benzoxazine monomers. In this case eugenol was first reacted with furfurylamine (Fig. 2.20) or stearylamine and then polymerized or copolymerized by ring opening polymerization. The resulting polymers were evidenced to exhibit very high char yield (up to 53.8 wt%) and very high LOI (up to 39%) depending on the comonomer content. The two latter examples show that fire properties may be enhanced by the introduction of aromatic monomer in their structure without addition of active elements like phosphorus or nitrogen.

2.3.4.3 Carbohydrate Derivatives

Ma et al. [115] used itaconic acid produced by fermentation of carbohydrates to synthesize phosphorus containing monomers with flame retardant activities. In a first step itaconic acid was reacted with DOPO in xylene at 120 °C to obtain DOPO containing itaconic acid (DI). In a second step two synthetic routes were envisaged to achieve the glycidylation of DI. Hence a DOPO containing monomer (EADI) with two epoxy functions was obtained (Fig. 2.21). This reactive flame retardant

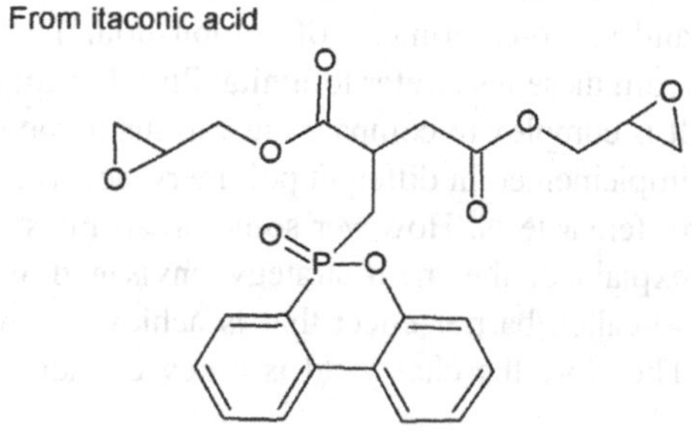

Fig. 2.21 Carbohydrate derivative based reactive flame retardant

From itaconic acid

was incorporated in an epoxy/anhydride resin. The fire behavior was assessed through UL94 vertical burning test and LOI. Even if the V-0 rating was not reached, the self-extinguishing time was considerably reduced as the phosphorus content increases. LOI increases from 19.6 to 31.4% as phosphorus increases from 0 to 2 wt%. When DGEBA monomer was completely replaced by EADI, the fire behavior was less performant. In this latter case, EADI strongly decreases the thermal stability of the resin that was not counterbalanced by the increase of char yield. In 2013, Kim et al. [15] used directly the alginic acid as polyol to react with 2,4-toluene di-isocyanate (TDI) for the synthesis of a polyurethane. Even if no fire properties were assessed, the thermal analysis of this alginate polyurethane hybrid reveals slow degradation kinetics and a high amount of residue at high temperature (circa 30 wt%) that seems promising with regard to flame retardancy.

2.4 Opportunities for the Industrial Scale-up of Biobased FR Systems

The preceding section has highlighted that a great number of research works has been carried out for the last few years to develop and assess new biobased flame retardants. Many biomolecules and green processes are available and each application is likely to find an appropriate fireproofing solution. Obviously, all the solutions studied for research purpose will not be scaled up and produced at industrial level. It is difficult to predict among the different proposals what products could be a success story in the coming years. However it seems that three main criteria could be considered: technical performance, health and environmental impact and economic efficiency.

2.4.1 Fire Performance Criteria

The first criterion to be considered is the fire performance. Do the new biobased flame retardants enable to fulfill the specifications of the fire regulation? In most of the scientific papers previously mentioned, four main types of tests have been carried out to characterize flammability properties: LOI, UL94, cone calorimeter and micro-calorimeter of combustion. The main parameters that can be extracted from these tests refer to ignitability, burning rate, char yield and self-extinguishing. It is complex to compare the various flame retardant systems because they are often implemented in different polymers, according to different processes and assessed by different tests. However some broad lines may be drawn up. As it was previously explained, the main strategy envisaged when using biobased compounds is the so-called barrier effect that is achieved through the promotion of a charred layer. Therefore the char yield is a key element of comparison. Char yield depends not only on the additive content but also on the presence of active element such as phosphorus in the formulation. Figures 2.15 and 2.17 show that char yield increases almost linearly with increasing phosphorus content in the material. Nevertheless the slope of the curve depends on the molecule bearing the phosphorus atoms and on the host matrix. Charring was generally related to lower pHRR and THR. Whatever the flame retardant and whatever the polymer, a linear decrease of pHRR with increasing char yield was evidenced as revealed by Figs. 2.11, 2.14, 2.16 and 2.18. A counterpart of charring is often an increase of ignability that results in a decrease of the onset of degradation (Fig. 2.13) and a decrease of the time-to ignition (TTI). In the case of biobased intrinsic flame retardant Fig. 2.12 shows that TTI tends to decrease with increasing content of additive. However some points were out of this trend. In the case of chitosan based LbL assemblies, it was highlighted in Fig. 2.14 that the temperature of degradation decreases with increasing char yield, i.e. with increasing additive content. Succeeding in decreasing the ignitability of polymers containing biobased flame retardant is very challenging for the coming years.

Apart fire performance, it is also important to check that the new biobased flame retardants do not impinge upon processability and other functional properties of polymers (e.g. mechanical properties, ageing resistance...).

2.4.2   Environmental and Health Criteria

For the last ten years some of the most widespread flame retardants have been phased out due to health and environmental issues. Some of the so-called halogenated compounds were proved to be bio-accumulative, some of them are considered to be endocrine disruptors, neurotoxic and carcinogenic. The development of new biobased flame retardants must of course avoid falling back into these issues. This is the objective of the REACH regulation and the role of the European Chemical Agency (ECHA) to promote the protection of human health and environment. Natural product does not mean safe product. Most of the biobased compounds mentioned in this review are harmless. However innocuousness must always be checked. As an example, some naturally found biophenols like catechol or hydroquinone may exhibited some toxicity [116]. Beside the intrinsic health and environmental properties of products, all the precepts of green chemistry must be favored when synthesizing new biobased flame retardants and a life cycle assessment approach should be used to determine the effective impact of these products.

2.4.3   Economic Criteria

Last but not least, economic criteria are probably those which could definitely impede the development of biobased flame retardants. A universal panacea would be low cost raw matter and low cost processes. With regard to the raw matters that were developed in this chapter, the most promising are probably the ones that derived from well-established industries. For example the wood industry supplies various biomolecules such as cellulose, lignin, lignosulfonate, vanillin that are of great interest for flame retardancy as described previously. Lignin can also be extracted from different other plants, especially in by-products. The world annual production of lignin is estimated to be higher than 50 million tons. Chitosan is another biomolecules presenting positive outlooks. There is a growing interest for crustacean shell-derived chitin and chitosan in biomedicine, nutrition and food processing. The global market for chitosan was estimated to be 21.4 thousand metric tons in 2015. A major issue impacting the chitosan is its high production cost. Nevertheless flame retardancy may take benefit of this growing market to constitute a new application of chitosan.

References

1. Anastas PT, Warner JC (1998) Green chemistry: theory and practice. Oxford University Press, Oxford
2. Biron M (2011) Biobased additives and their future. http://polymer-additives.specialchem.com/tech-library/article/bio-based-additives-their-future
3. Vassilev SV, Baxter D, Andersen LK, Vassileva CG (2010) An overview of the chemical composition of biomass. Fuel 89:913–933
4. Dubois J-L (2012) Refinery of the future: feedstock, processes, products. In: Aresta M, Dibenedetto A, Dumeignil F (eds) Biorefinery: from biomass to chemicals and fuels. De Gruyter, Berlin, pp 19–47
5. Richmond PA (1991) Occurence and functions of native cellulose. In: Haigler CH, Weimer PF (eds) Biosynthesis and biodegradation of cellulose. Marcel Dekker, Inc., New-York, pp 5–23
6. Shen DK, Gu S (2009) The mechanism for thermal decomposition of cellulose and its main products. Bioresour Technol 100:6496–6504
7. Dorez G, Ferry L, Sonnier R, Taguet A, Lopez-Cuesta JM (2014) Effect of cellulose, hemicellulose and lignin contents on pyrolysis and combustion of natural fibers. J Anal Appl Pyrolysis 107:323–331
8. Zobel HF (1988) Molecules to granules: a comprehensive starch review. Starch/Stärke 40:44–50
9. Liu X, Wang Y, Yu L, Tong Z, Chen L, Liu H, Li X (2013) Thermal degradation and stability of starch under different processing conditions. Starch/Staerke 65:48–60
10. Logithkumar R, Keshavnarayan A, Dhivya S, Chawla A, Saravanan S, Selvamurugan N (2016) A review of chitosan and its derivatives in bone tissue engineering. Carbohydr Polym 151:172–188
11. de Britto D, Campana-Filho SP (2007) Kinetics of the thermal degradation of chitosan. Thermochim Acta 465:73–82
12. Moussout H, Ahlafi H, Aazza M, Bourakhouadar M (2016) Kinetics and mechanism of the thermal degradation of biopolymers chitin and chitosan using thermogravimetric analysis. Polym Degrad Stab 130:1–9
13. McHugh DJ (2003) A guide to the seaweed industry, no. 441
14. Soares JP, Santos JE, Chierice GO, Cavalheiro ETG (2004) Thermal behavior of alginic acid and its sodium salt. Eclet Quim 29:57–63

15. Kim JS, Pathak TS, Yun JH, Kim KP, Park TJ, Kim Y, Paeng KJ (2013) Thermal degradation and kinetics of alginate polyurethane hybrid material prepared from alginic acid as a polyol. J Polym Environ 21:224–232
16. Anastasakis K, Ross AB, Jones JM (2011) Pyrolysis behaviour of the main carbohydrates of brown macro-algae. Fuel 90:598–607
17. Whiteford D (2005) Proteins: structure and functions. Wiley, New York
18. Moldoveanu SC (1998) Chapter 12: Analytical pyrolysis of proteins. In: Techniques and instrumentation in analytical chemistry: analytical pyrolysis of natural organic polymers, vol 20. pp 373-397
19. Mocanu AM, Moldoveanu C, Odochian L, Paius CM, Apostolescu N, Neculau R (2012) Study on the thermal behavior of casein under nitrogen and air atmosphere by means of the TG-FTIR technique. Thermochim Acta 546:120–126
20. Bates AD, Maxwell A (2005) DNA topology. Oxford University Press
21. Alongi J, Di Blasio A, Milnes J, Malucelli G, Bourbigot S, Kandola B, Camino G (2015) Thermal degradation of DNA, an all-in-one natural intumescent flame retardant. Polym Degrad Stab 113:110–118
22. Christie WW, Han X (2012) Lipid analysis: isolation, separation, identification and lipidomic analysis. Woodhead Publishing Ltd, Cambridge
23. Fahy E, Subramaniam S, Brown HA, Glass CK, Merrill AH, Murphy RC, Raetz CRH, Russell DW, Seyama Y, Shaw W, Shimizu T, Spener F, van Meer G, VanNieuwenhze MS, White SH, Witztum JL, Dennis EA (2005) A comprehensive classification system for lipids. J Lipid Res 46:839–861
24. Bedier AH, Hussein MF, Ismail EA, El-emary MM (2014) Jojoba and castor oils as fluids for the preparation of bio greases: a comparative study. Int J Sci Eng Res 5:755–762
25. The Oil Palm Wastes in Malaysia. World's Largest Science. Technology & Medicine Open Access Book Publisher
26. Gouveia De Souza A, Oliveira Santos JC, Conceição MM, Dantas Silva MC, Prasad S (2004) A thermoanalytic and kinetic study of sunflower oil. Braz J Chem Eng 21:265–273
27. Montero De Espinosa L, Meier MAR (2011) Plant oils: the perfect renewable resource for polymer science? Eur Polym J 47:837–852
28. Romani A, Lattanzio V, Quideau S (2014) Recent advances in polyphenol research, vol 4. Wiley Blackwell, Oxford
29. Brebu M, Vasile C (2010) Thermal degradation of lignin—a review. Cellul Chem Technol 44:353–363
30. Gaugler M, Grigsby WJ (2009) Thermal degradation of condensed tannins from radiata pine bark. J Wood Chem Technol 29:305–321
31. Camino G, Costa L, Luda di Cortemiglia MP (1991) Overview of fire retardant mechanisms. Polym Degrad Stab 33:131–154
32. Laoutid F, Bonnaud L, Alexandre M, Lopez-Cuesta JM, Dubois P (2009) New prospects in flame retardant polymer materials: From fundamentals to nanocomposites. Mater Sci Eng R Rep 63:100–125
33. Gallina G, Bravin E, Badalucco C, Audisio G, Armanini M, De Chirico A, Provasoli F (1998) Application of cone calorimeter for the assessment of class of flame retardants for polypropylene. Fire Mater 22:15–18
34. De Chirico A, Armanini M, Chini P, Cioccolo G, Provasoli F, Audisio G (2003) Flame retardants for polypropylene based on lignin. Polym Degrad Stab 79:139–145
35. Song P, Cao Z, Fu S, Fang Z, Wu Q, Ye J (2011) Thermal degradation and flame retardancy properties of ABS/lignin: Effects of lignin content and reactive compatibilization. Thermochim Acta 518:59–65
36. Ferry L, Dorez G, Taguet A, Otazaghine B, Lopez-Cuesta JM (2014) Chemical modification of lignin by phosphorus molecules to improve the fire behavior of polybutylene succinate. Polym Degrad Stab 113:135–143

37. Costes L, Laoutid F, Aguedo M, Richel A, Brohez S, Delvosalle C, Dubois P (2016) Phosphorus and nitrogen derivatization as efficient route for improvement of lignin flame retardant action in PLA. Eur Polym J 84:652–667

38. Alongi J, Cuttica F, Bourbigot S, Malucelli G (2015) Thermal and flame retardant properties of ethylene vinyl acetate copolymers containing deoxyribose nucleic acid or ammonium polyphosphate. J Therm Anal Calorim 122:705–715

39. Alongi J, Bosco F, Carosio F, Di Blasio A, Malucelli G (2014) A new era for flame retardant materials? Mater Today 17:152–153

40. Malucelli G, Bosco F, Alongi J, Carosio F, Di A (2014) Biomacromolecules as novel green flame retardant systems for textiles: an overview. RSC Adv 4:46024–46039

41. Bosco F, Carletto RA, Alongi J, Marmo L, Di Blasio A, Malucelli G (2013) Thermal stability and flame resistance of cotton fabrics treated with whey proteins. Carbohydr Polym 94:372–377

42. Alongi J, Carletto RA, Bosco F, Carosio F, Di Blasio A, Cuttica F, Antonucci V, Giordano M, Malucelli G (2014) Caseins and hydrophobins as novel green flame retardants for cotton fabrics. Polym Degrad Stab 99:111–117

43. Alongi J, Carletto RA, Di Blasio A, Carosio F, Bosco F, Malucelli G (2013) DNA: a novel, green, natural flame retardant and suppressant for cotton. J Mater Chem A 1:4779–4785

44. Alongi J, Cuttica F, Di Blasio A, Carosio F, Malucelli G (2014) Intumescent features of nucleic acids and proteins. Thermochim Acta 591:31–39

45. Alongi J, Milnes J, Malucelli G, Bourbigot S, Kandola B (2014) Thermal degradation of DNA-treated cotton fabrics under different heating conditions. J Anal Appl Pyrolysis 108:212–221

46. Alongi J, Carletto RA, Di Blasio A, Cuttica F, Carosio F, Bosco F, Malucelli G (2013) Intrinsic intumescent-like flame retardant properties of DNA-treated cotton fabrics. Carbohydr Polym 96:296–304

47. Bosco F, Casale A, Mollea C, Terlizzi ME, Gribaudo G, Alongi J, Malucelli G (2015) DNA coatings on cotton fabrics: effect of molecular size and pH on flame retardancy. Surf Coat Technol 272:86–95

48. Carosio F, Di Blasio A, Alongi J, Malucelli G (2013) Green DNA-based flame retardant coatings assembled through layer by layer. Polymer 54:5148–5153 (United Kingdom)

49. Carosio F, Cuttica F, Di Blasio A, Alongi J, Malucelli G (2015) Layer by layer assembly of flame retardant thin films on closed cell PET foams: efficiency of ammonium polyphosphate versus DNA. Polym Degrad Stab 113:189–196

50. Alongi J, Di Blasio A, Cuttica F, Carosio F, Malucelli G (2014) Bulk or surface treatments of ethylene vinyl acetate copolymers with DNA: Investigation on the flame retardant properties. Eur Polym J 51:112–119

51. Alongi J, Cuttica F, Carosio F, Applicata S, Torino P, Michel VT (2016) DNA coatings from byproducts: a panacea for the flame retardancy of EVA, PP, ABS, PET, and PA6? ACS Sustain Chem Eng 4:3544–3551

52. Réti C, Casetta M, Duquesne S, Bourbigot S, Delobel R (2008) Flammability properties of intumescent PLA starch and lignin. Polym Adv Technol 19:628–635

53. Wang X, Hu Y, Song L, Xuan S, Xing W, Bai Z, Lu H (2011) Flame retardancy and thermal degradation of intumescent flame retardant poly(lactic acid)/starch biocomposites. Ind Eng Chem Res 50:713–720

54. Dupretz R, Fontaine G, Bourbigot S (2013) Fire retardancy of a new polypropylene-grafted starch: Part II: investigation of mechanisms. J Fire Sci 32:210–229

55. Dupretz R, Fontaine G, Bourbigot S (2013) Fire retardancy of a new polypropylene-grafted starch. J Fire Sci 31:563–575

56. Carosio F, Fontaine G, Alongi J, Bourbigot S (2015) Starch-based layer by layer assembly: efficient and sustainable approach to cotton fire protection. ACS Appl Mater Interfaces 7:12158–12167

57. Carosio F, Alongi J, Malucelli G (2012) Layer by layer ammonium polyphosphate-based coatings for flame retardancy of polyester-cotton blends. Carbohydr Polym 88:1460–1469

58. Alongi J, Carosio F, Malucelli G (2012) Layer by layer complex architectures based on ammonium polyphosphate, chitosan and silica on polyester-cotton blends: Flammability and combustion behaviour. Cellulose 19:1041–1050

59. Jimenez M, Guin T, Bellayer S, Dupretz R, Bourbigot S, Grunlan JC (2016) Microintumescent mechanism of flame-retardant water-based chitosan-ammonium polyphosphate multilayer nanocoating on cotton fabric. J Appl Polym Sci 43783:1–12

60. Pan H, Song L, Ma L, Pan Y, Liew KM, Hu Y (2014) Layer-by-layer assembled thin films based on fully biobased polysaccharides: chitosan and phosphorylated cellulose for flame-retardant cotton fabric. Cellulose 21:2995–3006

61. Pan H, Wang W, Pan Y, Song L, Hu Y, Liew KM (2015) Formation of self-extinguishing flame retardant biobased coating on cotton fabrics via Layer-by-Layer assembly of chitin derivatives. Carbohydr Polym 115:516–524

62. Pan H, Song L, Hu Y, Liew KM (2015) An eco-friendly way to improve flame retardancy of cotton fabrics: layer-by-layer assembly of semi-biobased substance. Energy Procedia 75:174–179

63. Pan H, Wang W, Pan Y, Zeng W, Zhan J, Song L, Hu Y, Liew KM (2015) Construction of layer-by-layer assembled chitosan/titanate nanotubes based nanocoating on cotton fabrics: flame retardant performance and combustion behavior. Cellulose 22:911–923

64. Carosio F, Negrell-Guirao C, Alongi J, David G, Camino G (2015) All-polymer layer by layer coating as efficient solution to polyurethane foam flame retardancy. Eur Polym J 70:94–103

65. Pan Y, Zhan J, Pan H, Wang W, Tang G, Song L, Hu Y (2016) Effect of fully biobased coatings constructed via layer-by-layer assembly of chitosan and lignosulfonate on the thermal, flame retardant, and mechanical properties of flexible polyurethane foam. ACS Sustain Chem Eng 4:1431–1438

66. Zhang T, Yan H, Shen L, Fang Z, Zhang X, Wang J (2014) Chitosan/phytic acid polyelectrolyte complex: a green and renewable intumescent flame retardant system for ethylene–vinyl acetate copolymer. Ind Eng Chem Res 53:19199–19207

67. Zhang T, Yan H, Shen L, Fang Z, Zhang X, Wang J, Zhang B (2014) A phosphorus-, nitrogen- and carbon-containing polyelectrolyte complex: preparation, characterization and its flame retardant performance on polypropylene. RSC Adv 4:48285–48292

68. Alongi J, Pošsković M, Frache A, Trotta F (2010) Novel flame retardants containing cyclodextrin nanosponges and phosphorus compounds to enhance EVA combustion properties. Polym Degrad Stab 95:2093–2100

69. Enescu D, Alongi J, Frache A (2012) Evaluation of nonconventional additives as fire retardants on polyamide 6,6: phosphorous-based master batch, α-zirconium dihydrogen phosphate, and β-cyclodextrin based nanosponges. J Appl Polym Sci 123:3545–3555

70. Wang X, Xing W, Wang B, Wen P, Song L, Hu Y, Zhang P (2013) Comparative study on the effect of beta-cyclodextrin and polypseudorotaxane as carbon sources on the thermal stability and flame retardance of polylactic acid. Ind Eng Chem Res 52:3287–3294

71. Wang B, Qian X, Shi Y, Yu B, Hong N, Song L, Hu Y (2014) Cyclodextrin microencapsulated ammonium polyphosphate: preparation and its performance on the thermal, flame retardancy and mechanical properties of ethylene vinyl acetate copolymer. Compos Part B Eng 69:22–30

72. Qian W, Li XZ, Wu ZP, Liu YX, Fang CC, Meng W (2015) Formulation of intumescent flame retardant coatings containing natural-based tea saponin. J Agric Food Chem 63:2782–2788

73. Wang N, Hu L, Babu HV, Zhang J, Fang Q (2017) Effect of tea saponin-based intumescent flame retardant on thermal stability, mechanical property and flame retardancy of natural rubber composites. J Therm Anal Calorim 128:1133–1142

74. Cayla A, Rault F, Giraud S, Salaün F, Fierro V, Celzard A (2016) PLA with intumescent system containing lignin and ammonium polyphosphate for flame retardant textile. Polymers 8:331–346

75. Tondi G, Wieland S, Wimmer T, Thevenon MF, Pizzi A, Petutschnigg A (2012) Tannin-boron preservatives for wood buildings: mechanical and fire properties. Eur J Wood Wood Prod 70:689–696

76. Illy N, Fache M, Ménard R, Negrell C, Caillol S, David G (2015) Phosphorylation of bio-based compounds: the state of the art. Polym Chem 6:6257–6291

77. Aoki D, Nishio Y (2010) Phosphorylated cellulose propionate derivatives as thermoplastic flame resistant/retardant materials: Influence of regioselective phosphorylation on their thermal degradation behaviour. Cellulose 17:963–976

78. Pan H, Qian X, Ma L, Song L, Hu Y, Liew KM (2014) Preparation of a novel biobased flame retardant containing phosphorus and nitrogen and its performance on the flame retardancy and thermal stability of poly(vinyl alcohol). Polym Degrad Stab 106:47–53

79. Costes L, Laoutid F, Khelifa F, Rose G, Brohez S, Delvosalle C, Dubois P (2016) Cellulose/phosphorus combinations for sustainable fire retarded polylactide. Eur Polym J 74:218–228

80. Božič M, Liu P, Mathew AP, Kokol V (2014) Enzymatic phosphorylation of cellulose nanofibers to new highly-ions adsorbing, flame-retardant and hydroxyapatite-growth induced natural nanoparticles. Cellulose 21:2713–2726

81. Hu S, Song L, Hu Y (2013) Preparation and characterization of chitosan-based flame retardant and its thermal and combustible behavior on polyvinyl alcohol. Polym Plast Technol Eng 52:393–399

82. Hu S, Song L, Pan H, Hu Y (2013) Effect of a novel chitosan-based flame retardant on thermal and flammability properties of polyvinyl alcohol. J Therm Anal Calorim 112:859–864

83. Hu S, Song L, Pan H, Hu Y (2012) Thermal properties and combustion behaviors of chitosan based flame retardant combining phosphorus and nickel. Ind Eng Chem Res 51:3663–3669

84. Hu S, Song L, Pan H, Hu Y, Gong X (2012) Thermal properties and combustion behaviors of flame retarded epoxy acrylate with a chitosan based flame retardant containing phosphorus and acrylate structure. J Anal Appl Pyrolysis 97:109–115

85. Howell BA, Carter KE, Dangalle H (2011) Flame retardants based on tartaric acid: a renewable by-product of the wine industry. ACS Symposium Series, chapter 9, pp 133–152

86. Howell BA, Carter KE (2010) Thermal stability of phosphinated diethyl tartrate. J Therm Anal Calorim 102:493–498

87. Howell BA, Daniel YG (2015) Thermal degradation of phosphorus esters derived from isosorbide and 10-undecenoic acid. J Therm Anal Calorim 121:411–419

88. Mauldin TC, Zammarano M, Gilman J, Shields JR, Boday D (2014) Synthesis and characterization of isosorbide-based polyphosphonates as biobased flame-retardants. Polym Chem 5:5139–5146

89. Liu Y, Wang JS, Zhu P, Zhao JC, Zhang CJ, Guo Y, Cui L (2016) Thermal degradation properties of biobased iron alginate film. J Anal Appl Pyrolysis 119:87–96

90. Liu Y, Zhao XR, Peng YL, Wang D, Yang L, Peng H, Zhu P, Wang DY (2016) Effect of reactive time on flame retardancy and thermal degradation behavior of bio-based zinc alginate film. Polym Degrad Stab 127:20–31

91. Zhang J, Ji Q, Shen X, Xia Y, Tan L, Kong Q (2011) Pyrolysis products and thermal degradation mechanism of intrinsically flame-retardant calcium alginate fibre. Polym Degrad Stab 96:936–942

92. Liu Y, Zhang CJ, Zhao JC, Guo Y, Zhu P, Wang DY (2016) Bio-based barium alginate film: preparation, flame retardancy and thermal degradation behavior. Carbohydr Polym 139:106–114

93. Costes L, Laoutid F, Dumazert L, Lopez-cuesta J-M, Brohez S, Delvosalle C, Dubois P (2015) Metallic phytates as efficient bio-based phosphorous flame retardant additives for poly(lactic acid). Polym Degrad Stab 119:217–227

94. Yu Y, Fu S, Song P, Luo X, Jin Y, Lu F, Wu Q, Ye J (2012) Functionalized lignin by grafting phosphorus-nitrogen improves the thermal stability and flame retardancy of polypropylene. Polym Degrad Stab 97:541–546

95. Prieur B, Meub M, Wittemann M, Klein R, Bellayer S, Fontaine G, Bourbigot S (2015) Phosphorylation of lignin to flame retard acrylonitrile butadiene styrene (ABS). Polym Degrad Stab 127:32–43

96. Liu L, Huang G, Song P, Yu Y, Fu S (2016) Converting industrial alkali lignin to biobased functional additives for improving fire behavior and smoke suppression of polybutylene succinate. ACS Sustain Chem Eng 4:4732–4742

97. Liu L, Qian M, Song P, Huang G, Yu Y, Fu S (2016) Fabrication of green lignin-based flame retardants for enhancing the thermal and fire retardancy properties of polypropylene/wood composites. ACS Sustain Chem Eng 4:2422–2431

98. Zhang R, Xiao X, Tai Q, Huang H, Hu Y (2012) Modification of lignin and its application as char agent in intumescent flame-retardant poly(lactic acid). Polym Eng Sci 52:2620–2626

99. Zhang R, Xiao X, Tai Q, Huang H, Yang J, Hu Y (2012) Preparation of lignin–silica hybrids and its application in intumescent flame-retardant poly(lactic acid) system. High Perform Polym 24:738–746

100. Marosi G, Toldy A, Parlagh G, Nagy Z, Ludányi K, Anna P, Keglevich G (2002) A study on the selective phosphorylation and phosphinylation of hydroxyphenols. Heteroat Chem 13:126–130

101. Vothi H, Nguyen C, Lee K, Kim J (2010) Thermal stability and flame retardancy of novel phloroglucinol based organo phosphorus compound. Polym Degrad Stab 95:1092–1098

102. Ménard R, Negrell-Guirao C, Ferry L, Sonnier R, David G (2014) Synthesis of biobased phosphate flame retardants. Pure Appl Chem 86:1637–1650

103. Ménard R, Negrell C, Fache M, Ferry L, Sonnier R, David G (2015) From a bio-based phosphorus-containing epoxy monomer to fully bio-based flame-retardant thermosets. RSC Adv 5:70856–70867

104. Ménard R, Negrell C, Ferry L, Sonnier R, David G (2015) Synthesis of biobased phosphorus-containing flame retardants for epoxy thermosets comparison of additive and reactive approaches. Polym Degrad Stab 120:300–312

105. Vahabi H, Sonnier R, Ferry L (2015) Effects of ageing on the fire behaviour of flame-retarded polymers: a review. Polym Int 64:313–328

106. Lligadas G, Callau L, Ronda JC, Galià M, Cádiz V (2005) Novel organic-inorganic hybrid materials from renewable resources: hydrosilylation of fatty acid derivatives. J Polym Sci Part A: Polym Chem 43:6295–6307

107. Lligadas G, Ronda JC, Gali M, Cdiz V, Galia M, Ca V (2006) Novel silicon-containing polyurethanes from vegetable oils as renewable resources. Synth Prop Biomacromolecules 7:2420–2426

108. Lligadas G, Ronda JC, Galià M, Cádiz V (2006) Synthesis and properties of thermosetting polymers from a phosphorous-containing fatty acid derivative. J Polym Sci Part A: Polym Chem 44:5630–5644

109. Lligadas G, Ronda JC, Galià M, Cádiz V (2006) Development of novel phosphorus-containing epoxy resins from renewable resources. J Polym Sci Part A: Polym Chem 44:6717–6727

110. Heinen M, Gerbase AE, Petzhold CL (2014) Vegetable oil-based rigid polyurethanes and phosphorylated flame-retardants derived from epoxydized soybean oil. Polym Degrad Stab 108:76–86

111. Zhang L, Zhang M, Hu L, Zhou Y (2014) Synthesis of rigid polyurethane foams with castor oil-based flame retardant polyols. Ind Crops Prod 52:380–388

112. Pillai CKS, Prasad VS, Sudha JD, Bera SC, Menon ARR (1990) Polymeric resins from renewable resources. II. Synthesis and characterization of flame-retardant prepolymers from cardanol. J Appl Polym Sci 41:2487–2501

113. Wan J, Gan B, Li C, Molina-Aldareguia J, Li Z, Wang X, Wang D-Y (2015) A novel biobased epoxy resin with high mechanical stiffness and low flammability: synthesis, characterization and properties. J Mater Chem A 3:21907–21921
114. Thirukumaran P, Shakila Parveen A, Sarojadevi M (2014) Synthesis and copolymerization of fully biobased benzoxazines from renewable resources. ACS Sustain Chem Eng 2:2790–2801
115. Ma S, Liu X, Jiang Y, Fan L, Feng J, Zhu J (2014) Synthesis and properties of phosphorus-containing bio-based epoxy resin from itaconic acid. Sci China Chem 57:379–388
116. Michałowicz J, Duda W (2007) Phenols—Sources and toxicity. Polish J Environ Stud 16:347–362

Chapter 3
Flame Retardancy of Natural Fibers Reinforced Composites

Due to numerous advantages (high specific mechanical properties, low density, biosourcing, ...), natural fibers from plants are considered as credible alternatives to glass or carbon fibers for composites industry. Nevertheless, their relatively high flammability limits their potential applications. Many researches have been carried out to improve the flame retardancy of composites reinforced with natural fibers. This chapter attempts to establish the state-of-art of these researches.

There are already some reviews about the flammability of natural fibers and biocomposites [1–5]. Compared to these reviews, the specific objective of this chapter is to answer to the following questions: What are the differences in flammability between composites filled natural fibers and their counterparts reinforced with glass or carbon fibers? Is the treatment of natural fibers a viable solution for flame retarding biocomposites? The chapter is based on the literature available on biocomposites filled with natural fibers. The articles about the flammability of composites based on wood flour are not taken into account but reviews about this topic can be found elsewhere [6].

3.1 A Comparison Between Natural Fibers and Glass or Carbon Fibers

3.1.1 Fire Behavior of Composites Filled with Glass or Carbon Fibers

Since natural fibers should replace glass (GF) or carbon fibers (CF), it is important to list main issues specific to these latter. Due to their high stability, glass and carbon fibers do not directly participate to the heat release during burning. Carbon fibers start degrading only at high heat flux in cone calorimeter in presence of air, i.e. at the end of the test when flame is vanished [7]. Nevertheless, their presence

© The Author(s) 2018
R. Sonnier et al., *Towards Bio-based Flame Retardant Polymers*,
Biobased Polymers, DOI 10.1007/978-3-319-67083-6_3

has an important influence (and sometimes negative influence) on the flammability. In particular, glass fibers lead generally to a decrease in limiting oxygen index for thermoplastic composites [8–10]. Casu et al. have explored the reasons for such a phenomenon [11]. They showed that the decrease in limiting oxygen index of glass fibers reinforced composites is not due to the seizing of fibers but mainly to the anti-dripping effect of glass fibers: fibers prevent the flowing of burning polymers away from the flame. Seizing may be another influent parameter [12]. Additional effect is due to wicking. Wicking corresponds to the flowing of polymer along the glass fiber surface from the bulk to the fire zone. Interfacial charring by grafting flame retardants on glass fiber surface can prevent efficiently the wicking [13]. Glass fibers also prevent dripping during UL94 test, another common fire test assessing the flammability of a small sample [14].

On the contrary, glass or carbon fibers have a positive effect in forced-flaming tests, for example in cone calorimeter test [7, 9–11, 14, 15]. Indeed, flowing is prevented in cone calorimeter test. Hence, the replacement of combustible by inert fibers decreases the fire load. High thermal conductivity of fibers allows delaying ignition [14, 15]. Perret et al. observed an increase of 20–25 s of time-to-ignition for epoxy reinforced with carbon fibers in cone calorimeter test at 35 kW/m^2 [15]. Liu et al. have measured the physical properties of flame retarded polystyrene containing various contents of glass fibers [14]. They showed that the incorporation of glass fibers enhances the thermal conductivity and the density but decreases the specific heat capacity. The net effect is an increase of the thermal inertia which influences the time-to-ignition [16]. Nevertheless Dao et al. noted that a small increase of carbon fiber in epoxy composites leads to a worse thermal resistance (decrease of time-to-ignition) but no explanation was proposed to explain this tendency [7].

A high thermal inertia generally delays ignition but increases the heating rate in the deeper layers. This was observed by Liu et al. in the case of polystyrene-based composites reinforced with glass fibers [14]. Usually, the fast heating of the bulk results in a higher heat release rate in a second time [17]. Nevertheless such a phenomenon is not observed for composites. Indeed, the accumulation of fibers at the top surface of samples promotes the formation of an insulating residue able to limit the gases and heat transfer, even if high thermal conductive fibers are probably not the best candidates for insulation. Consequently the heat release rate decreases.

Fibers can also modify more or less the modes-of-action of flame retardants. For example, Camino et al. noted that intumescence is hindered by the presence of carbon fibers in epoxy composites [18]. Braun et al. suggested that the impact of polymer charring promotion on heat release rate is limited in composites filled with high loadings of fibers [19]. Some authors have reported that the residue content of composites reinforced with glass or carbon fibers is higher than expected from a linear rule of mixtures. Kandola and Toqueer-Ul-Haq assumed that the presence of glass fibers allows increasing slightly char content in thermogravimetric analysis for PP flame retarded with some phosphate-based molecules [9]. They explained that glass fibers act as physical barrier and provide more chance for fire retardants to react with polymer. Zhao et al. [20] have incorporated small amount of carbon fibers (up to 5 wt%) into thermoplastic polyurethane (TPU) flame retarded with ammonium polyphosphate (APP). They also noted that carbon fibers lead to a significant increase of char yield in thermogravimetric analysis: from 21 wt% for TPU-APP to 30–33 wt% for TPU-APP-CF. Carbon fibers also promote a dense and compact char in cone calorimeter test. Liu et al. noted that glass fibers incorporated into flame retarded polystyrene leads to higher char content than expected in cone calorimeter tests but not in thermogravimetric analysis [14]. The high residue content in cone calorimeter test may be probably assigned to the insulating character of the inert glass fibers accumulated at the top surface and limiting the degradation of the underlying polymer.

Orientation and aspect ratio of fibers in the composite are also important parameters influencing heat conductivity and flammability. Chai et al. have compared the flammability of epoxy composites containing different glass fabrics at the same volume fraction [21]. They pointed out significant differences in fire performances, for example in pHRR at cone calorimeter test. Levchik et al. showed that the limiting oxygen index of epoxy resin reinforced with carbon fibers is much higher if the fibers are oriented perpendicularly to the front of flame [18]. Milled glass fibers reduce the oxygen index of poly(butylene terephthalate) at a lower extent than long glass fibers because wicking effect was limited [11]. Kandola and Toqueer-Ul-Haq observed that the residue content of polypropylene filled with glass fibers in cone calorimeter tests was much higher when the formulations were not compounded using extrusion but prepared directly by hot pressing [9]. Indeed, despite the non-uniform distribution of the fibers in the non-compounded composite filled with 20 wt% of glass fibers, the residue content was 29.5 wt% versus 19.1 wt% for the extruded formulation. The authors explained that the glass fibers were longer when not compounded and formed a better physical barrier at the top surface.

Finally, it must be noted that a small variation of fibers can modify greatly the fire performance of the composite. The change in fire properties is not linear with fiber content. Dao et al. [7] have studied epoxy composites reinforced with 56 or 59 vol.% of carbon fibers in cone calorimeter. The composite containing the highest amount of carbon fibers exhibits lower time-to-ignition but also lower mass loss rate and lower mass loss at the end of test. In particular at low heat flux (20 kW/m^2, probably close to critical heat flux) the mass loss at the end of test is twice lower.

3.1.2 Flammability of Natural Fibers

Natural fibers share some characteristics with glass or carbon fibers but exhibit also some important differences. Surely, they increase the viscosity and should prevent the flowing of polymers as do glass and carbon fibers. Similarly, intumescence should be limited in the presence of natural fibers. On the contrary, natural fibers may not promote high heat transfer. Thermal conductivity of natural fibers is close to that of polymers and much lower than glass or carbon fibers [3]. Annie Paul et al.

have measured the thermal conductivity of PP composites filled with banana fibers [22]. They found that the thermal conductivity decreases when incorporating 50 vol.% of fibers: from 0.240 to 0.157 W/m K. Similar results were found about polyester resins filled with sisal and banana fibers [23]. In both studies, various pretreatments of fibers allowed increasing the thermal conductivity due to a better compatibility with the matrix. Idicula et al. have also studied a composite with pineapple leaf fiber (PALF) and/or glass fibers [23]. They were able to estimate the thermal conductivity of the PALF fibers: 0.188 W/m K, i.e. much less than glass fibers (around 1–1.2 W/m K). Consequently, natural fibers do not allow transferring heat from the surface to the bulk. Insulating properties of the residue should be better than with high thermal conductive glass fibers. Note that some authors have claimed surprising and contradictory results: Du et al. observed a significant increase of thermal conductivity for PP filled with ramie fibers (from 0.22 to 0.41 W/m K with 27.5 wt% of ramie fibers [24].

Obviously, the most relevant difference between natural fibers and glass or carbon fibers is their thermal decomposition. Natural fibers are ligno-cellulosic materials and then their flammability does not depend only on cellulose (the main component), but also on hemicellulose and lignin. The composition of a wide range of natural fibers can be found elsewhere [3, 5]. These three components have different thermal stability. Hemicellulose is the least thermally stable and starts decomposing at 200 °C. Cellulose is decomposing around 350 °C. Lignin is decomposing on a wide range of temperatures from 200 to 600 °C. Moreover the activation energy of pyrolysis is very different as recalled by Yao et al. [25]: 105–111, 195–213 and 35–65 kJ/mol, for hemicellulose, cellulose and lignin, respectively. The decomposition steps are largely overlapping.

The decomposition of cellulose, the main component in natural fibers, has been extensively studied. Briefly, after initial desorption of water, cellulose decomposes through two competitive routes. The first one corresponds to the formation of levoglucosan by depolymerisation. Then levoglucosan decomposes itself in high flammable low molecular weight molecules and char. It is well known that the formation of this levoglucosan during degradation of natural fibers leads to worse thermal and fire properties. The second route is the dehydration of cellulose giving more char [1, 4, 26].

An important point is that natural fibers are able to char. This char allows forming a barrier layer preventing the transfer of the heat from the flame to the underlying polymer. Even if the intrinsic char yield remains limited in some cases, it can be significantly enhanced by the use of additives, particularly phosphorus-based flame retardants. As polyaromatic compound, lignin is the main component promoting the fiber charring. Then its influence on flammability deserves to be discussed in greater details. As already noted, lignin is a minor component of natural fibers but its content varies in a large range (from 0 wt% for pure cellulose, less than 3 wt% for flax fibers [27], 20–30 wt% for bamboo fibers [28, 29] and up to 45–48 wt% for piassava fibers [30]). Dorez et al. have studied the properties of various natural fibers considering them as a mixture of three main components: cellulose, hemicellulose (i.e. xylan) and lignin [31]. Char yield increases when lignin content increases but strong discrepancy can be found between experimental values and char yield calculated from a linear rule of mixtures. In particular, at low lignin content, experimental char yield is much higher than expected. The authors propose that lignin degradation releases acids able to dehydrate cellulose and to promote its charring. Lignin also increases the activation energy of combustion. Nevertheless other studies have already noted an opposite effect: lignin would prevent the thermal polymerization of levoglucosan, leading to a reduction of char fraction [32]. Yao et al. have studied a wide range of natural fibers by thermogravimetric analysis. They did not find any significant correlation between lignin content and char yield or apparent activation energy of pyrolysis (around 150–175 kJ/mol) [25]. To explain the apparently contradictory results about the role of lignin, it can be assumed that the composition and the structure of the fibers are much complex and may not be oversimplified using a linear rule of mixtures on the lignin content. Indeed, Lewin et al. have shown for a long time the influence of fine structure of cellulose (crystallinity, orientation, molecular weight, interchain distance and hydrogen bonds) on pyrolysis [33, 34]. It is obvious that the even more complex structure of natural fibers must also play a significant role on pyrolysis.

The heat released from the combustion of natural fibers releases around 8–12 kJ per gram of gases, which should contribute to the propagation of the flame [31, 35]. Nevertheless, this contribution is limited. Indeed, taking into account the relatively low effective heat of combustion and the char yield of many fibers, heat released by the burning of fibers is only around 8 kJ per gram of fibers. This is much lower than many polymers as PE, PP, PBS or PLA. Therefore the incorporation of natural fibers leads to a decrease in total heat release of many composites (Fig. 3.1). Of course, in the case of low flammable polymers like polyfurfuryl alcohol [36] or polybenzoxazine [37], the contribution of natural fibers to heat release becomes prominent.

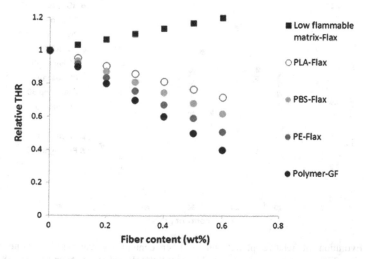

Fig. 3.1 Calculated relative total heat release of composites based on various polymers containing glass fibers or flax fibers—values are calculated according to a linear rule of mixtures

3.2   Fire Properties of Composites Reinforced by Natural Fibers

3.2.1   Influence of Raw Natural Fibers

There are now consistent results about the influence of natural fibers on flammability of biocomposites. Natural fibers include at least flax, hemp, bamboo, banana, kenaf, sugar cane. Polymer matrices are mainly PE, PP, PLA and PBS. Attempts were made to incorporate natural fibers into engineering plastics as PA6 and PBT [37]. In all cases, natural fibers lead to a decrease of peak of heat release rate [35, 38–40] (Fig. 3.2). For a given series of composites, the decrease seems to be proportional to the fiber content. The best results indicate a decrease of pHRR around 14% for each addition of 10 wt% of fibers [35, 38]. One reference shows a slightly higher decrease of pHRR when 25 wt% of flax fibers are incorporated into a blend of PLA and thermoplastic starch (PLA/TPS) plasticized by glycerol [41]. A minimum content of natural fibers is required to form a protective barrier against fire. For example, Dorez et al. [35] evidenced that over 10 wt% of flax fibers, the HRR curves exhibit a plateau, typical of a barrier effect.

As expected from already discussed considerations, time-to-ignition is often reduced due to the presence of low thermally stable fibers [35, 38]. Nevertheless, this decrease is not systematic or sometimes is not significant [39–41].

Char yield of composites is quite often increasing when incorporating natural fibers, in particular when polymer is unable to char as polyolefins or aliphatic polyesters. In most articles, the experimental char yields are not compared to those calculated using a linear rule of mixtures. In the case of PBS filled with different

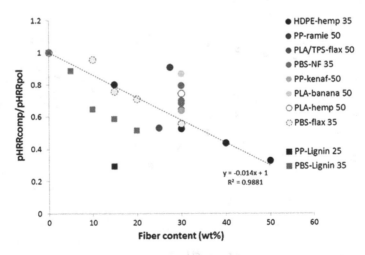

Fig. 3.2 Evolution of relative pHRR versus fiber content for various non-flame retarded biocomposites. Figures on caption correspond to heat flux (plotted results from [24, 35, 38–41, 55, 56, 63, 65])

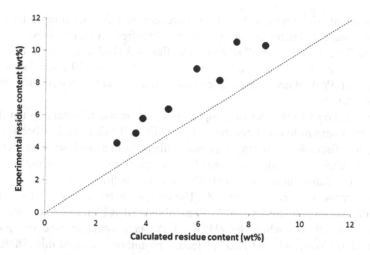

Fig. 3.3 Experimental versus calculated residue contents of biocomposites PBS-flax fibers in anaerobic pyrolysis (from TGA results [35])

contents of various natural fibers, Dorez et al. observed the experimental char yield is systematically slightly higher than the expected values [35] (Fig. 3.3).

Apart from cone calorimeter, other fire characterizations were made on various biocomposites. We report here only results about the most common fire tests.

LOI is often slightly reduced in the presence of natural fibers [41–43] but it is not systematic [24, 44]. For example, Sain et al. have observed a slight increase in LOI for PP filled with 50 wt% of sawdust or rice husk (from 24 to 25–26) [45]. The decrease in LOI is assigned by some authors to the wicking effect [41, 43, 46]. This explanation is discussed in a further section. When a matrix is not rated at UL94 vertical burning test, the incorporation of natural fibers does not allow improving the rating [40–42, 47, 48]. Dripping is still observed in many cases. Glow wire flammability index of a PP copolymer drops from 725 to 675 °C when adding 30 wt% of kenaf fibers [40]. This last value is in good agreement with that found by Schartel et al. for PP filled with 30 wt% of flax fibers (650 °C) [47]. Horizontal burning rates (ASTM D 635) are significantly increasing after incorporation of 50 wt% of sawdust or rice husk into PP [45]. Jeencham et al. also observed an increase of horizontal burning rate for PP containing 30 wt% of sisal fibers using the same test [42]. This is not surprising considering the effect of a low thermal stability on propagation rate predicted by some models [49]. On the contrary, Suardana et al. noted a decrease in burning rate at the same test for PLA and PP after incorporation of jute or coconut fibers (volume fraction 35%) [50].

It is noteworthy that beyond the common tendencies listed just above, strong differences can be found depending on the type of natural fibers. There are relatively few studies comparing the fire performances of biocomposites (i.e. one polymer reinforced with various natural fibers). For example, Dorez et al. compared PBS composites filled with pure cellulose, flax, hemp, sugar cane and bamboo at a weight content of 30 wt% [35]. The differences in terms of time-to-ignition or pHRR in cone calorimeter test are significant. Manfredi et al. observed even more variations for acrylic resins filled with jute, flax and sisal fibers (30% in volume) [51]. Time-to-ignition in cone calorimeter at 35 kW/m^2 is 110 s for flax but only 52 s for sisal. Peak of heat release rate is close to 500 and 900 kW/m^2 for sisal and jute, respectively.

Note that the performances of composites based on natural fibers do not depend only on the composition and content of fibers. Even if this point has been scarcely studied, the dispersion and length of short fibers in thermoplastics, or the fabric structure in thermoset composites must have a significant impact. Wang et al. have compared the flammability of two HDPE composites filled with 15 wt% of hemp short fibers using cone calorimeter [38]. The composite containing 2 wt% of maleic anhydride-grafted polyethylene (MAPE) exhibits a HRR plateau and a lower pHRR (around 2000 kW/m^2) while the pHRR of the non-compatibilized composite is close to 2750 kW/m^2 without HRR plateau. The authors assigned this difference to the enhanced dispersion of fibers into the matrix due to MAPE allowing a more efficient barrier effect. Chai et al. have compared three epoxy composites based on flax fabrics with different structures using various fire tests [21]. For example, in cone calorimeter, the performances of the three composites are different in terms of time-to-ignition or peak of heat release rate. In vertical burning test, the burning duration and the mass loss also change according to the fabric structure.

3.2.2 Role of Lignin

We have already discussed about the role of lignin on the flammability of natural fibers. Its influence on the fire behavior of biocomposites was highlighted by several studies.

Shukor et al. have studied biocomposites based on PLA and kenaf fibers using APP [52]. They compared raw and alkali treated kenaf fibers. Alkali treatment using different NaOH concentration allows removing lignin. Lignin content is not mentioned in this work but it is around 8–19 wt% in raw kenaf fibers [28, 53]. The thermal stability of composites is slightly enhanced when lignin is removed. But residue content and LOI are reduced. LOI decreases from 29.4 when raw kenaf fibers are used to 28.0 when fibers are treated with the highest concentration of NaOH. Similarly, residue content in anaerobic pyrolysis at 600 °C decreases from 19.4 to 16 wt%. Jute fibers were added to vinylester resin [54]. The decomposition temperature of both components (jute and resin) tends to increase by a few degrees (2–5 °C) after alkaline treatment of fibers (which reduces hemicellulose and lignin contents). But contrarily to Shukor's work, the residue content of the composite increases significantly when jute fibers are treated.

Due to the high char yield of lignin, several authors have attempted to use it directly alone or in combination with other flame retardants [55–58]. De Chirico et al. have incorporated 15 wt% of lignin into non-charring polypropylene [55]. The peak of heat release rate is reduced from 1400 to 410 kW/m^2. Such a drop (−70%) is much more impressive than for composites filled with natural fibers (around −20, −25% at the same content). Ferry et al. have also observed a significant decrease in pHRR for PBS composites containing various contents of alkali lignin (−40% for 15 wt% of lignin) [56] (Fig. 3.2). In De Chirico's work, the formation of char is only due to lignin (15 wt% of lignin leading to 6 wt% of char) but it is enough to observe a strong decrease of pHRR. Chen et al. have also observed that the char yield of composites PP filled with alkylated or raw kraft lignin increases when more lignin is incorporated [59]. Nevertheless the char yields are always below the values calculated from a linear rule of mixtures (Fig. 3.4). On the contrary, Ferry et al. have concluded about some interactions between lignin and PBS because the experimental char contents in thermogravimetric analysis are slightly higher (around +2 wt%) than the values calculated from a linear rule of mixtures [56]. Actually, the explanation may be more complicated as the experimental char contents of PBS filled with lignin-free fibers (pure cellulose) are also higher than expected [35].

Different types of lignin exist and exhibit various compositions and charring capacity. De Chirico et al. measured a char content of 40 wt% at 700 °C [55] while Ferry et al. noted that alkali and organosolv lignin exhibit a char yield of 53.5 and 58.7 wt% at 750 °C, respectively [56]. Moreover, the latter authors have attempted to enhance the char yield by grafting phosphorus molecules onto lignin. No significant increase in char yield was noted. Moreover, the fire behavior of PBS containing 20 wt% of lignin is unchanged in cone calorimeter after modifying lignin. But the authors note that the residues of PBS composites with modified lignin are more homogeneous and cohesive with fewer cracks.

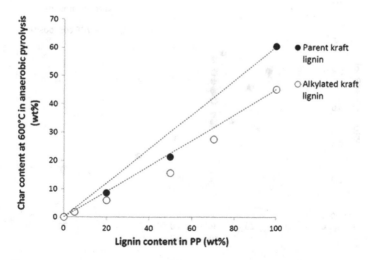

Fig. 3.4 Char content in anaerobic pyrolysis versus lignin content for PP-lignin biocomposites (from [59]). *Dotted lines* represent the values calculated from a linear rule of mixtures

Most generally, the incorporation of lignin into a polymer leads to a reduction of thermal stability and time-to-ignition [55, 56]. Nevertheless it is not always the case. Hence, Chen et al. have studied PP composites filled with kraft lignin [59]. The temperatures for 10 and 50 wt% of mass loss increase from 406 to 443 °C for pure PP to 433 and 472 °C for PP containing 5 wt% of alkylated lignin. At higher content, the degradation starts earlier.

Zhang et al. pointed out another effect of lignin on flammability of composites [57]. The authors have prepared flame retarded silicones filled with high amount of lignin. The best ranking at UL94 test (V-0) can be reached if thermal and chemical post-treatments are applied to remove Si-H groups on silicones. They observed that the decomposition rate of composites is slowed down above 450 °C and explained that lignin acts as phenolic antioxidant trapping radicals and then protecting silicone from radical degradation.

The presence of lignin may also change the capacity of natural fibers to be grafted by some molecules as revealed by Dorez et al. [60] and discussed in the following.

3.2.3   Comparison with CF or GF Composites

When the influence of natural fibers is compared to that of glass or carbon fibers in terms of pHRR reduction, it seems that glass or carbon fibers may perform slightly better (Fig. 3.5). But a systematic comparison rather than a compilation of results from various articles would give more reliable conclusions.

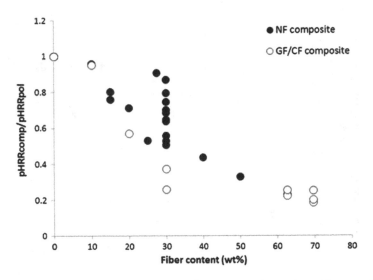

Fig. 3.5 Evolution of relative pHRR versus fiber content for various biocomposites and composites (plotted results from [9–11, 15, 24, 35, 39–41, 63, 65])

Some articles deal with the comparison of the flammability of composites reinforced with glass fibers or alternatively natural fibers. But systematic comparison is scarce. El Sabbagh et al. have studied PA6 and PBT composites filled with flax or glass fibers using a cone calorimeter [61]. Some composites were flame retarded with organic phosphate and/or boehmite. The fiber content was fixed at 30 wt%. There is no clear tendency when comparing similar composites reinforced with flax or glass fibers. In some cases, the glass fibers-reinforced composite performs better (lower pHRR). But similar results are obtained in several cases. Non-flame retarded PA6 composite filled with flax fibers exhibit a lower pHRR than its counterpart reinforced by glass fibers. Same conclusions may be drawn considering THR.

Same results can be drawn from the work of Hapuarachchi et al. [62]. The authors have compared flammability of sheet moulding compounds (SMC) reinforced with hemp or glass fibers using cone calorimeter at various heat fluxes. Some SMC were flame retarded with aluminum trihydrate. SMC filled with hemp fibers exhibit higher THR than SMC filled with glass fibers with or without flame retardant. pHRR was lower for non-flame retarded SMC filled with glass fibers. But when aluminum trihydrate was added, SMC reinforced with hemp fibers performed better. More surprisingly, SMC filled with hemp fibers ignited later than their counterparts, particularly at low heat flux (25 kW/m^2).

Glass fibers are denser than natural fibers. Then when composites reinforced with glass or natural fibers are compared at the same weight loading of fibers, the amount of matrix is higher in the composite reinforced with glass fibers. Another (more valuable?) comparison is to study composites filled with the same volume fraction of fibers. Hence the amount of matrix is kept constant. This was the choice of Chai et al. [21]. The authors have compared different epoxy composites based on flax and glass fabrics. The volume fraction of fibers was fixed at 40%. Characterizations include horizontal and vertical burning tests and cone calorimeter tests. The structure of glass fabrics is similar to that of flax fabrics. Results show clearly that glass fibers-reinforced composites perform better. They exhibit higher time-to-ignition, lower peak of heat release rate, lower mass loss (and consequently lower total heat release). Indeed, at same volume fraction, the same amount of polymer can burn but flax fibers also contribute to heat release in composites filled with natural fibers. Moreover, the authors noted that glass-reinforced composites maintain their initial shape. On the contrary flax fibers-reinforced composites deform with delamination and bulging. Hapuarachchi and Peijs have also reported such a phenomenon [63]. Manfredi et al. have compared various composites based on unsaturated polyester or acrylic resin and reinforced with natural or glass fibers at a same volume fraction (30%) [51]. Based on cone calorimeter test, the authors confirmed that glass fibers-filled composites exhibit better performances, and in particular a lower heat release as expected.

As noted above, some authors consider that natural fibers promote wicking effect as glass fibers. Nevertheless, it is doubtful that the flowing of polymer is favored along the natural fibers. Indeed, natural fibers have a lower thermal conductivity and are much rougher than smooth glass fibers. Moreover, they are able to char even without flame retardant. In other words, the interfacial charring proposed by some authors for composites filled with glass fibers [13] can improve the flame retardancy of biocomposites by reducing the heat release and increasing the char yield but not by preventing the unrelevant wicking effect.

3.3 Flame Retardancy of Biocomposites

In most cases, the flame retardancy of natural fibers reinforced composites was provided by the incorporation of flame retardant additives into the matrix. In many articles it is not specified if the addition of the FR system is expected to have an effect on the natural fibers or on the matrix thermal degradation pathways or on both. Another strategy could be to modify specifically the degradation pathway of NF by grafting or adsorbing or encapsulating the fibers with appropriate species (molecules or particles). The modification of the chemistry of the fiber is often performed in order to avoid the formation of levoglucosan [64].

3.3.1 Incorporation of Flame Retardants into the Matrix

The most used FR additive is surely ammonium polyphosphate, which is well known as char promoter in presence of a carbonization agent [35, 47, 48, 52, 65, 66]. While natural fibers are rich in polyhydric compounds, ammonium polyphosphate can be very effective even in composites based on polyolefin matrix as polypropylene for example. In that case, natural fibers can be considered as a part of a multicomponent FR system. Other phosphorus compounds as 9,10-dihydro-9-oxa-10-phosphaphenanthrene-10-oxide (DOPO), phosphoric acid or dihydrogen ammonium can act similarly to APP [67, 68].

Le Bras et al. have prepared PP/flax fibers composites flame retarded with APP and with a complex intumescent system (IFR—based on APP, melamine and pentaerythritol) [66]. The ratio PP/fibers is fixed to 1.5. The authors observed that both flame retarded composites perform similarly in cone calorimeter test. In other words, APP and flax react to form a charred layer protecting the underlying material similarly to an intumescent system, confirming that flax can be used as a charring source. Nevertheless, the PP/flax/APP composite is not rated at UL94 test (too long combustion time and formation of burning drops) while the PP/flax/IFR composite is V-0 rated. Schartel et al. [47] have also observed that APP and flax react to form a charred layer. Moreover, APP allows to increase the LOI from 21 (only PP with 30% of flax) to 26 (when 25% of PP was replaced by APP). A V-0 rated was obtained by Nie et al. [48] using 20% of microencapsulated APP (MCAPP) in a PBS/bamboo fiber (30/50) composite. Similarly, Reti et al. tried to replace the typical pentaerythritol (PER) with 10% of lignin (replacement of petroleum sources with natural renewable resources) used as charring source into intumescent flame

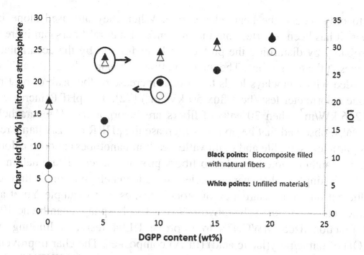

Fig. 3.6 Influence of diglycidylphenylphosphate on the flame retardancy of materials based on DGEBA and natural fibers (from [44])

retarded PLA (containing 30 wt% APP) [69]. Lignin does not perform as well as PER in terms of LOI and pHRR in cone calorimeter. On the contrary, lignin allows reaching V0 rating at UL94 test while the formulation with PER is only V2 rated.

It is noteworthy that the combination of phosphorus flame retardants and natural fibers does not lead systematically to a synergy. For example, Sudhakara et al. have studied composites based on DGEBA and Borassus fruit fibers [44]. The composites were flame retarded with various contents of diglycidylphenylphosphate (DGPP). This flame retardant improves the flame retardancy of the composites, in terms of char yield, LOI or burning rate. But the improvement is the same with or without 5 wt% of fibers (Fig. 3.6). This may be assigned to the low content of fibers.

Other flame retardants have been used. A detailed list of flame retardants can be found elsewhere [1, 5]. Sain et al. used magnesium dihydroxide and boron-based compounds to improve the flame retardancy of PP filled with sawdust and rice husk [45]. Hapuarachchi et al. incorporated mineral filler (aluminum trihydrate) into SMC filled with hemp fibers [62]. Jeencham et al. have attempted to flame retard PP/sisal using zinc borate or magnesium hydroxide [42]. But when mineral flame retardants and phosphorus compounds have been compared at same contents, it seems that the former were not as efficient as the latter [42, 48, 70]. It may be due to the fact that mineral fillers like magnesium or aluminum hydroxide are efficient only at very high loadings. But the main reason is probably that natural fibers and phosphorus compounds cooperate as already explained: phosphorus flame retardant is the acid source reacting with the hydroxyl groups of the fiber leading to dehydration and charring of cellulose [71].

Nanoparticles are scarcely used alone to improve fire retardancy because they are not able to improve all the aspects of the fire behavior. Indeed, they are mainly efficient to only reduce the heat release rate. When they are used alone in bio-composites, it has been reported that the presence of natural fibers can increase the pHRR probably by disturbing the protective layer formed by the accumulation of nanoparticles at the top surface. The incorporation of banana fibers into a PP matrix flame retarded with nanoclays leads to a linear increase of the peak of heat release rate in cone calorimeter test (heat flux 50 kW/m^2) [72]. The pHRR increases from 748 to 1256 kW/m^2 when 30 wt% of fibers are incorporated. Hapuarachchi and Peijs have also observed that hemp fibers increase the pHRR of PLA flame retarded with nanoparticles (sepiolite and multiwalled carbon nanotubes) in cone calorimeter test [63]. The authors noted that hemp fibers provoke severe delamination during the test. Interestingly, when nanoparticles are adequately modified, they can be highly efficient on flame retardancy of biocomposites. For example Yu et al. [68] proved the efficiency of 9,10-dihydro-9-oxa-10-phosphaphenanthrene-10-oxide (DOPO) functionalized MWCNTs to improve UL94 test and limiting oxygen index (LOI) of ramie/poly(lactic acid) (PLA) composites. The char improvement is allowed thanks to a high dispersion of the functionalized MWCNTs and the interaction between DOPO and ramie fibers (that acts as a charring agent).

On the whole, phosphorous compounds seem to be the most efficient FR in NF based biocomposites. As they are incorporated by additive way, a high amount is usually needed that can induce a significant degradation of mechanical properties and may provoke processing troubles [61, 73]. For example, Zhang et al. have studied PP composites filled with APP and silica. Both fillers lead to a decrease in mechanical properties, particularly impact strength. Moreover, the authors note that these fillers have a certain impact during foaming using CO_2 as blowing agent [73]. One alternative that can be used for laminates is to protect the composite by a fire retardant layer (multilayer composites) as proposed by several authors [74, 75]. The geometry as well the choice of the composition for each layer has a great influence on properties, including fire properties [74, 76]. Another alternative is to incorporate the FR in reactive way, in order to flame retard the reinforcement. Therefore, the question is: Is it possible to reach the same fire performances by flame retarding only the natural fibers? A less ambitious (but more realistic) objective may be to combine both approaches in order to reduce the FR amount in the matrix: the incorporation of FR additives into the matrix and the chemical grafting of FR on the natural fibers.

3.3.2 Is to Flame Retard Natural Fibers Enough to Impart Fire Performances to Composites?

Natural fibers are sometimes functionalized to improve their compatibility with the matrix. Even if these treatments are not intended to improve the flame retardancy, some authors have noted that the fiber modifications have an impact on the fire behavior, especially on LOI [77–80]. Hence, Nair et al. [78] found that PS composites filled with modified (benzoylated and acetylated) sisal fibers exhibit a better thermal stability than the composite containing unmodified fibers. According to the authors, this is explained by the improved stability of modified fibers themselves but also by the interactions between the modified fibers and the matrix. Silanated Grewia optiva fibers increase the thermal stability and the residue content of unsaturated polyester filled with 30 wt% of fibers [77]. The limiting oxygen index is also slightly enhanced from 19 to 22 when fibers are silanated. VP et al. [39] have not observed that silanization of banana fibers does improve the thermal stability of PLA biocomposites. But the burning rate is reduced in UL-94 horizontal test, dripping is limited and drips do not ignite cotton. Moreover the pHRR in cone calorimeter decreases from 398 kW/m^2 for PLA containing 30 wt% of untreated banana fibers to 340 kW/m^2 for PLA filled with silanated fibers at the same loading.

The strategies to flame retard natural reinforcements can be directly inspired from techniques used in textile industry. There is a huge literature about the flame retardancy of plant-based textiles, mainly cotton. Alongi et al. [81] have recently reviewed the innovative methods to impart flame retardancy to fabrics. Among the reviewed methods, layer-by-layer (LbL) assembly process has attracted the interest of several researchers. Flame retardancy was imparted mainly to cotton fabrics [82–85], but also ramie fabrics [86] or delignified wood fibers [87]. In most cases, phosphorus-based compounds were used to promote the charring. To the best of our knowledge, LbL has rarely been used to flame retard a natural reinforcement for composite applications. Li et al. [37] have coated a ramie fabric using this method and incorporated it into a polybenzoxazine composite. The coating is constituted by ammonium polyphosphate (APP) and polyethyleneimine. The fabric content was around 65–70 wt%. The LbL treatment strongly increases the LOI: from 26.5 for the composite containing the uncoated fabric to 38.9. The coating also allows reaching V-0 rating at UL94 test. The thermal stability is enhanced above 300 °C under nitrogen and air and the residue content is significantly increased (by 15% under nitrogen). In this case, due to the high amount of ramie fibers and to the intrinsic good behavior of the matrix, the flame retardancy of fibers allows dramatically improving the performances of the composite.

Some authors have modified natural fibers by adsorbing or absorbing some phosphorous flame retardants as DAP, APP or phosphoric acid without washing fibers after modification [43, 46, 50]. For example, Suardana et al. have treated coconut and jute fibers with diammonium phosphate and incorporated them into PLA or PP at a volume fraction of 35% [50]. The phosphorus content into fibers was not measured. The flammability of composites is reduced as proved by the decrease in burning rate (horizontal propagation) and the increase in char yield.

Grafting of FR is another strategy. In most cases, it is involved through hydroxyl groups of cellulose or lignin. Dorez et al. used molecules (cellulose, xylan and lignin) to study the reactivity of natural fibers toward phosphonic acid molecules (such as octadecylphosphonic acid, ODPA) [60]. They proved that ODPA reacts strongly with lignin and at a lower degree with xylan (model for hemicellulose) whereas no reaction with cellulose was evidenced. These results mean that in soft conditions used in this study, phosphonic acid preferentially reacts with aromatic hydroxyl groups rather than with aliphatic ones. Then the modification may be much more effective for lignin-rich natural fibers.

Du et al. have chemically wrapped via in situ condensation reaction a phosphorus and nitrogen containing flame retardant onto ramie fibers [24]. The authors have firstly carried out an alkali treatment on fibers to "activate" hydroxyl groups, i.e. to remove hemicellulose and lignin but also surface impurities (other treatments may be used as ethanol to remove waxes and pectins ensuring the cohesion of fibers beam [67]). The phosphorus content was not precised but the char yield under nitrogen increased from 8 to more than 40 wt% for the fibers. Fibers were incorporated at 40 phr into PP (containing a small part of maleated polypropylene). The char yield of composites was 0 wt% for PP and PP filled with unmodified ramie fibers, but 12.6 wt% for PP filled with phosphorylated fibers. As expected, the thermal stability of modified fibers was reduced leading to a lower time-to-ignition in cone calorimeter test. The peak of heat release rate decreased from 714 kW/m^2 for PP containing unmodified ramie fibers to 548 kW/m^2 when fibers were phosphorylated. LOI was also slightly enhanced (from 18 to 20).

Dorez et al. have compared the phosphorylation of flax fibers using molecular (dihydrogen ammonium phosphate—DAP) or macromolecular (poly(methacryloyloxy)methyl phosphonic acid homopolymer and poly(methacryloyloxy)methyl phosphonic acid methylmethacrylate copolymer) flame retardants [88]. For similar phosphorus content, thermal stability is more reduced and charring is more enhanced with DAP. According to the authors, it is due to the fact that phosphorus from DAP is covalently directly bonded to flax while only a fraction of phosphorus from macromolecular flame retardants is involved in the grafting. In other words, the phosphorus fraction not directly bonded to flax is less efficient to promote dehydration of cellulose (maybe because a less close proximity). When the authors incorporate the phosphorylated flax fibers onto PBS matrix, they observe that the degradation of PBS is not affected by the presence of fibers. Phosphorus acts as char promoter only for the fibers and not for the matrix. It seems that DAP does not perform better than macromolecular flame retardants when composites are tested using cone calorimeter. The authors explain that these phosphonated macromolecules are able to char themselves leading to the formation of a sheath that acts as a barrier effect. This effect is not highlighted in thermogravimetric analysis because barrier effect is not efficient at microscale.

While phosphorus directly bonded to flax fibers seems to be more effective to promote charring, the same authors have compared in deeper details the grafting of octadecylphosphonic acid (ODPA) and dimethyl(octadecyl)phosphonate on the different components of flax [60]. The authors have shown that only phosphonic acid function is able to be grafted and not phosphonate group.

Natural fibers differ from synthetic fibers by their high porosity. Therefore the natural fibers can be flame retarded in bulk and high amounts of FR can be grafted while synthetic fibers are flame retarded only on their surface (except if a flame retardant is added directly during processing). Nevertheless, in many articles, the location of flame retardants onto (or into) the fibers is not specified. This point has been addressed by Sonnier et al. [27]. These authors have used ionizing radiations to graft phosphorus flame retardants into flax fibers. Ionizing radiations allow grafting FR to ligno-cellulosic materials without involving hydroxyl groups. The main drawback of this method is that cellulose is highly sensitive to irradiation and its molecular weight decreases drastically even at moderate dose (above 20 kGy). Nevertheless the authors have shown that it is possible to graft high amounts of phosphorus onto flax fibers at low doses (more than 3 wt% of phosphorus at 10 kGy) when molecules (dimethylvinyl phosphonate—MVP) diffuse into the fiber bulk. On the contrary the grafting content is low (<0.5 wt% of phosphorus) when the flame retardant (dimethyl(methacryloxy)methyl phosphonate—MAPC1) remains on the fiber surface. In this work, it was not possible to determine if flame retardants were grafted to cellulose or homopolymerized and trapped into the fiber core. Self-extinguishing flax fabrics are obtained when the phosphorus content exceeds 1 wt%. Char content is roughly proportional to the phosphorus content. When modified flax fabrics are incorporated into unsaturated polyester, the fire behavior of the composite is slightly improved (using cone calorimeter test) (Fig. 3.7). Nevertheless, phosphorous flame retardants act mainly as char promoter and then do not modify the pyrolysis of the matrix. It results only on the charring of the fabric (representing around 30 wt% of the total weight of the composite). Then this strategy does not allow the composite to reach the highest level of flame retardancy.

If the flame retardancy of natural fibers is most often carried out by using phosphorous flame retardants, other possibilities exist. As an example, Zhang et al. have flame retarded cellulose fibers by grafting zinc ion [20]. Maleic anhydride is first grafted on cellulose by reacting with hydroxyl groups. Then zinc ions were grafted onto modified fibers using zinc carbonate. Strong decrease of flammability is measured using cone calorimeter and LOI. LOI is improved from 19 to 30 when zinc content increases to 4.96 wt%. The authors consider that zinc ion catalyzes the

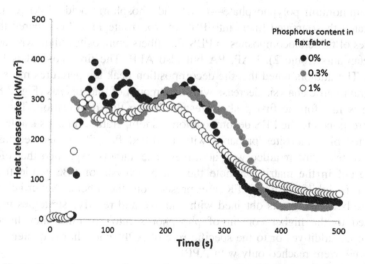

Fig. 3.7 Heat release rate curves of composites containing flax fabrics (unpublished data)

dehydration of cellulose, inhibiting its depolymerization and promoting its charring. Moreover, metal compounds are generated, covering the fibers surface to stop the oxygen penetration and the release of pyrolytic gases.

Combined strategies may also be considered. Zhou et al. have coated kenaf fibers with zinc oxide [89]. The objective of these authors is to partially replace a phosphorus-based flame retardant (resorcinol di(phenyl phosphate)—RDP) by these modified fibers into flame retarded PLA. The ZnO coating allows RDP to be adsorbed onto kenaf fibers. Therefore, the dispersion of fibers is improved and a better flame retardant effect is claimed by the authors. Indeed, PLA is V-0 rated at UL94 test with only 9 wt% of RDP and 6 wt% of modified kenaf fibers. Nevertheless, no comparison was performed with similar composites containing unmodified fibers and thermogravimetric analyses do not support any enhancement of char formation.

Composites based on natural fibers have become attracted for ten or fifteen years, i.e. more or less when halogenated flame retardants are gradually replaced by other alternatives. Then there are very few studies about the use of halogenated compounds to flame retard biocomposites. Misra et al. have brominated coir fibers to improve the LOI of epoxy composites [90]. Nevertheless, the coir content is only 5 wt% and fibers are used as filler rather than as reinforcement.

3.3.3   Comparison of Both Approaches

Some researchers have compared both approaches cited above, i.e. the incorporation of FR additives and the functionalization of natural reinforcement to flame retard biocomposites.

Dorez et al. have treated flax fibers with two different flame retardants (dihydrogen ammonium polyphosphate—DAP and phosphoric acid—PA) [67]. They incorporated these treated fibers into PBS matrix (route 1) and compared the performances of these biocomposites to PBS-flax fibers composites filled with additive flame retardants (route 2): DAP, PA but also APP. The fiber content is fixed at 30 wt%. The authors noted that the decomposition peak temperatures measured in thermogravimetric analysis decrease with the amount of phosphorus (Fig. 3.8). The decrease is fast for the first peak corresponding to flax degradation. The second peak corresponds to the PBS decomposition. Its temperature remains stable around 400 °C for biocomposites prepared with modified flax fibers but decreases for biocomposites flame retarded with additives. The authors explained that FR additives present in the matrix promote the hot hydrolysis of PBS while this phenomenon is inhibited when FR are present on the fibers. Nevertheless, the difference between results obtained with additive and reactive strategies may also be related to the higher content of phosphorus into biocomposites filled with phosphorous additives or to the specific role of APP (phosphorus contents higher than 1 wt% were reached only with APP).

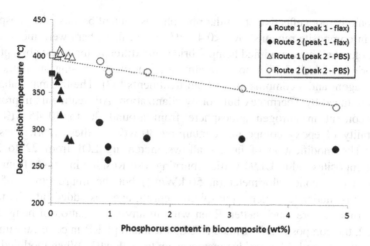

Fig. 3.8 Decomposition temperatures of PBS and flax fibers in biocomposites PBS-flax fibers (from [67])

Flammability of biocomposites was assessed using cone calorimeter at 35 kW/m² (Fig. 3.9). DAP on fibers is more effective to promote charring than into PBS matrix. Nevertheless, only the incorporation of APP at higher contents allows reaching high char yields (up to 30 wt%). The peak of heat release rate decreases when phosphorus content increases. But this tendency is not linear. On the one hand, the incorporation of additives into PBS matrix leads to lower pHRR because this route leads to higher phosphorus contents. But on the other hand the proximity of the phosphorous compound with the flax fibers may allow the formation of an efficient barrier layer even at low phosphorus content. It is not possible to determine

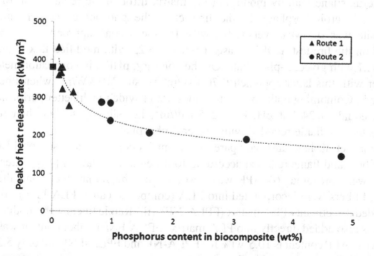

Fig. 3.9 Peak of heat release rate (in cone calorimeter) of biocomposites PBS-flax fibers (from [67])

which route is more effective at similar phosphorus content because the phosphorus content in composite remains low (<0.4 wt%) when flax fibers were modified.

Szolnoki et al. have modified hemp fabric using different methods: absorption of phosphoric acid (called Thermotex treatment), modification using an aminosilane coupling agent and a combination of both treatments [43]. The thermal stability of fibers is reduced by Thermotex but not by silanization. All treatments improve the residue content in nitrogen atmosphere from around 30 to 40–45 wt%. The flammability of epoxy composites containing 30 wt% of these fabrics has been studied. The modification of fabrics allows increasing LOI from 22 to 26–28 but all composites fail to UL94 vertical burning test. Residue increases from 1.8 to 7.4–9.6 wt% in cone calorimeter (at 50 kW/m^2) but the improvement of flame retardancy is moderate. When a phosphorus curing agent is added into the matrix, the performances are much better. Even without any modification of hemp fabric, LOI is 32, the composite is V-1 rated at UL94 test and pHRR in cone calorimeter is significantly reduced. Char residue increases up to 20.1 wt%. When modified fabric is combined with flame retarded epoxy resin, LOI remains equal to 32 but V-0 rating is reached. Residue content in cone calorimeter is not further improved (around 20 wt%) but peak of heat release rate is reduced and delayed. Then, the phosphorylation of the matrix is much efficient than the modification of fabric. This result can be basically related to the dispersion of phosphorus (present in matrix bulk and not only on the fabric) but also to the phosphorus content. When only fabric is modified, the phosphorus content in the biocomposite does not exceed 0.5 wt%. Biocomposite based on flame retarded epoxy is 1.75 wt% even without any phosphorylation of the fabric.

Bocz et al. have studied the flammability of various PLA/TPS plasticized with glycerol and containing 25 wt% of flax fibers [41]. In their study and contrarily to the previous ones, the phosphorus content is higher by modifying the fibers using a phosphorus silane than by modifying the matrix through the replacement of glycerol by glycerol-phosphate. In the first case, the phosphorus content into the composite was 0.5 wt% (versus 0.2 wt% for the second approach). The results depend on the fire test. LOI increases from 20 to 23 with modified flax fibers, and only to 21 with glycerol-phosphate. On the contrary, pHRR in cone calorimeter test is lower with this latter approach (270 kW/m^2 versus 310 kW/m^2 when fibers are modified). Combining both approaches does not provide much better performances. LOI is equal to 24 but pHRR is 285 kW/m^2, i.e. slightly higher than for the composite only flame retarded with glycerol-phosphate.

Shumao et al. have also compared different biocomposites based on PLA and ramie fibers and flame retarded according to different approaches [46]. Three composites were prepared: (i) APP was adsorbed on the ramie fibers and then the modified fibers were incorporated into PLA (composite called PLA-FNF). (ii) APP was added directly into PLA matrix (FPLA-NF). (iii) Ramie fibers were modified and APP was also added directly into PLA matrix (FPLA-FNF). Fiber content was fixed at 30 wt%. APP content is 10.5 wt% for FPLA-NF and FPLA-FNF but only 5.3 wt% for PLA-FNF composite. Therefore the latter composite has poor fire performances: LOI increases from 19.1 (for unmodified composite) to 25 and no rating is reached at

UL94 test. Char residue under air increases from 8.9 to 13.3 wt%. Both other composites (with 10.5 wt% of APP) perform much better. Residue content reaches 19.5–21.9 wt% and V-0 rating is obtained. Moreover FPLA-FNF allows reaching a LOI value of 35.6 versus only 28.1 for FPLA-NF. These results show that it is more efficient to flame retard the fibers in addition to the matrix. According to the authors, it may be due to the candlewick effect which is inhibited when the fibers are treated but we have already discussed this assumption. Note that mechanical tests on bio-composites confirm that modification of fibers allow maintaining better mechanical performances. This may be related to the presence of APP in matrix for FPLA-NF and FPLA-FNF which deteriorates the mechanical properties but the authors also noted from SEM observations that interfacial adhesion between PLA and fibers is better when APP is adsorbed onto ramie.

The main issue of these studies is that both approaches (flame retarded matrix or flame retarded fibers) are compared while the phosphorus content is not the same. With the exception of the study of Bocz et al. [41], the phosphorus content is always higher when an additive phosphorous flame retardant is added into the matrix. This is explained by the fact that the phosphorus content provided by this last method can be (almost) as high as desired. On the contrary, the phosphorus content provided by the modification of fibers is often limited. However, as already discussed, it exists some works allowing to graft high phosphorus content (up to 4 wt%) into a natural reinforcement [27].

3.4 Conclusions

Biocomposites filled with natural fibers are nowadays very attractive. In order to extend their application fields, their fire behavior needs to be studied. A quite large but recent literature has been focused on the flame retardancy of these biocom-posites. From the results reviewed above, it is possible to draw some main con-clusions, even if some specific points are still controversial.

Despite their combustible nature, natural fibers have been very soon considered by researchers as a potential component of flame retardant systems due to their capacity to charring in the presence of phosphorus additives. Ammonium polyphosphate is used in a considerable number of research works to reach flame retarded biocomposites.

Nevertheless, FR additives impact negatively mechanical properties and pro-cessing of biocomposites as for all polymeric materials. Therefore, efforts have been made to provide flame retardancy through the modification of the natural reinforcement. From this point of view, the porosity and the surface functions of natural fibers offer more opportunities than synthetic fibers. This strategy has also involved mainly phosphorous flame retardants.

Flame retarded reinforcements have been incorporated into composites. Even if the flammability is reduced, the highest level of flame retardancy cannot be reached without adding flame retardants into the matrix. The main reason of such limited performances is twofold. First, the phosphorus content grafted onto fibers remains most often limited. Second, the flame retardants are located only on the reinforcement and not dispersed into the matrix. As explained above, most of articles dealt with charring promoters (based on phosphorus) to improve the flame retardancy of natural fibers. In such a case, flame retardants act only on the fibers but the matrix (representing generally more than 50% of the weight) remains unaffected. Due to their high porosity, it should be possible to incorporate high amounts of flame inhibitors able to slowdown the combustion of pyrolytic gases released from the degradation of fibers but also of the matrix. This strategy alone or in combination with charring promoters may be more efficient.

Other aspects deserve to be further investigated. Let us quote two among them. There is very little knowledge about the influence of the fiber structure on the grafting efficiency. At best, the grafting was studied on the separated main components of the fibers, i.e. cellulose, hemicellulose and lignin but the organization of all the components including minor ones can be very complex. The formation of the barrier char layer when natural fibers are decomposing is a main parameter to reach high flame retardancy. Not only the composition and the content of the fibers must be taken into account but also the fabric structure, or in case of thermoplastic matrix filled with short fibers, or the length, entanglement and orientation of fibers. Only few studies have provided some data about these issues.

References

1. Mngomezulu M, John M, Jacobs V, Luyt A (2014) Review on flammability of biofibres and biocomposites. Carbohyd Polym 111:149–182
2. Prabhakar M, Rehaman Shah A, Song J-I (2015) A review on the flammability and flame retardant properties of natural fibers and polymer matrix based composites. Compos Res 28:29–39
3. Kozlowski R, Wladyka-Przybylak M (2008) Flammability and fire resistance of composites reinforced by natural fibers. Polym Adv Technol 19:446–453
4. Azwa Z, Yousif B, Manalo A, Karunasena W (2013) A review on the degradability of polymeric composites based on natural fibres. Mater Des 47:424–442
5. Chapple S, Anandjiwala R (2010) Flammability of natural fiber-reinforced composites and strategies for fire retardancy: a review. J Thermoplast Compos Mater 23:871–893
6. Kim J, Pal K (2010) Flammability in WPC composites, chapter 6. In: Recent advances in the processing of wood-plastic composites. Engineering materials, vol 32. Springer, pp 129–147
7. Dao D, Rogaume T, Luche J, Richard F, Bustamante Valencia L, Ruban S (2016) Thermal degradation of epoxy resin/carbon fiber composites: influence of carbon fiber fraction on the fire reaction properties and on the gaseous species release. Fire Mater 40:27–47
8. Casu A, Camino G, Luda M, De Giorgi M (1993) Mechanisms of fire retardance in glass fibre polymer composites. Makromol Chem Macromol Symp 74:307–310
9. Kandola B, Toqueer-Ul-Haq R (2012) The effect of fibre content on the thermal and fire performance of polypropylene-glass composites. Fire Mater 36:603–613
10. Köppl T, Brehme S, Wolff-Fabris F, Alstädt V, Schartel B, Döring M (2012) Structure-property relationships of halogen-free flame-retarded poly(butylene terephthalate) and glass fiber reinforced PBT. J Appl Polym Sci 124:9–18

11. Casu A, Camino G, De Giorgi M, Flath D, Laudi A, Morone V (1998) Effect of glass fibres and fire retardant on the combustion behaviour of composites, glass fibres-poly(butylene terephtalate). Fire Mater 22:7–14

12. Vahabi H, Lopez-Cuesta JM, Chivas C (2017) High performance flame retardant polyamide materials, chap 6. In: De-Yi Wang (ed) Novel flame retardant polymers and composite materials. Woodhead Publishing

13. Liu L, Liu Y, Han Y, Liu Y, Wang Q (2015) Interfacial charring method to overcome the wicking action in glass fiber-reinforced polypropylene composite. Compos Sci Technol 121:9–15

14. Liu J, Guo Y, Zhang Y, Liu H, Peng S, Pan B, Ma J, Niu Q (2016) Thermal conduction and fire property of glass fiber-reinforced high impact polystyrene/magnesium hydroxide/microencapsulated red phosphorus composite. Polym Degrad Stab 129:180–191

15. Perret B, Schartel B, Stöb K, Ciesielski M, Diederichs J, Döring M, Krämer J, Altstädt V (2011) A New Halogen-free flame retardant based on 9,10-Dihydro-9-oxa-10-phosphaphenanthrene-10-oxide for epoxy resins and their carbon fiber composites for the automotive and aviation industries. Macromol Mater Eng 296:14–30

16. Hopkins D, Quintiere J (1996) Material fire properties and predictions for thermoplastics. Fire Saf J 26:241–268

17. Patel P, Hull R, Stec A, Lyon R (2011) Influence of physical properties on polymer flammability in the cone calorimeter. Polym Adv Technol 22:1100–1107

18. Levchik S, Camino G, Costa L, Luda M (1996) Mechanistic study of thermal behavior and combustion performance of carbon fibre-epoxy resin composites fire retarded with a phosphorus-based curing system. Polym Degrad Stab 54:317–322

19. Braun U, Balabanovich A, Schartel B, Knoll U, Artner J, Ciesielski M, Döring M, Perez R, Sandler J, Altstädt V, Hoffmann T, Pospiech D (2006) Influence of the oxidation state of phosphorus on the decomposition and fire behavior of flame-retarded epoxy resin composites. Polymer 47:8495–8508

20. Zhang K, Zong L, Tan Y, Ji Q, Yun W, Shi R, Xia Y (2016) Improve the flame retardancy of cellulose fibers by grafting zinc ion. Carbohyd Polym 136:121–127

21. Chai M, Bickerton S, Bhattacharyya D, Das R (2012) Influence of natural fibre reinforcements on the flammability of bio-derived composite materials. Compos B 43:2867–2874

22. Annie Paul S, Boudenne A, Ibos L, Candau Y, Joseph K, Thomas S (2008) Effect of fiber loading and chemical treatments on thermophysical properties of banana fiber/polypropylene commingled composite materials. Compos A 39:1582–1588

23. Idicula M, Boudenne A, Umadevi L, Ibos L, Candau Y, Thomas S (2006) Thermophysical properties of natural fibre reinforced polyester composites. Compos Sci Technol 66:2719–2725

24. Du S-I, Lin X-B, Jian R-K, Deng C, Wang Y-Z (2015) Flame-retardant wrapped ramie fibers towards suppressing "candlewick effect" of Polypropylene/Ramie fiber composites. Chin J Polym Sci 33:84–94

25. Yao F, Wu Q, Lei Y, Guo W, Xu Y (2008) Thermal decomposition kinetics of natural fibers: activation energy with dynamic thermogravimetric analysis. Polym Degrad Stab 93:90–98

26. Horrocks A (1983) An introduction to the burning behaviour of cellulosic fibres. J Soc Dye Colour 99:191–197

27. Sonnier R, Otazaghine B, Viretto A, Apolinario G, Ienny P (2015) Improving the flame retardancy of flax fabrics by radiation grafting of phosphorus compounds. Eur Polym J 68:313–325

28. Müssig J, Fischer H, Graupner N, Drieling A (2010) Testing methods for measuring physical and mechanical fibre properties (plant and animal fibres). In: Müssig J (ed) Industrial applications of natural fibres: structure, properties and technical applications. Wiley, New York, pp 269–309

29. Wahab R, Mustafa M, Sudin M, Mohamed A, Rahman S, Samsi H, Khalid I (2013) Extractives, Holocellulose, α-Cellulose, Lignin and Ash Contents in Cultivated Tropical Bamboo Gigantochloa brang, G. levis, G. scortechinii and G. wrayi. J Biol Sci 5:266–272

30. John M, Anandjiwala R (2008) Recent developments in chemical modification and characterization of natural fiber-reinforced composites. Polym Compos 29:187–207
31. Dorez G, Ferry L, Sonnier R, Taguet A, Lopez-Cuesta J-M (2014) Effect of cellulose, hemicellulose and lignin contents on pyrolysis and combustion of natural fibers. J Anal Appl Pyrol 107:323–331
32. Hosoya T, Kawamoto H, Saka S (2007) Cellulose-hemicellulose and cellulose-lignin interactions in wood pyrolysis at gasification temperature. J Anal Appl Pyrol 80:118–125
33. Rodrig H, Basch A, Lewin M (1975) Crosslinking and pyrolytic behavior of natural and man-made cellulosic fibers. J Poly Sci 13:1921–1932
34. Lewin M (2005) Unsolved problems and unanswered questions in flame retardance of polymers. Polym Degrad Stab 88:13–19
35. Dorez G, Taguet A, Ferry L, Lopez-Cuesta J-M (2013) Thermal and fire behavior of natural fibers/PBS biocomposites. Polym Degrad Stab 98:87–95
36. Kumar R, Anandjiwala R (2013) Compression-moulded flax fabric-reinforced polyfurfuryl alcohol bio-composites. J Therm Anal Calorim 112:755–760
37. Li N, Yan H, Xia L, Mao L, Fang Z, Song Y, Wang H (2015) Flame retarding and reinforcing modification of ramie/polybenzoxazine composites by surface treatment of ramie fabric. Compos Sci Technol 121:82–88
38. Wang K, Addiego F, Laachachi A, Kaouache B, Bahlouli N, Toniazzo V, Ruch D (2014) Dynamic behavior and flame retardancy of HDPE/hemp short fiber composites: effect of coupling agent and fiber loading. Compos Struct 113:74–82
39. Mohanty S, Nayak S (2015) A study on thermal degradation kinetics and flammability properties of poly(lactic acid)/banana fiber/nanoclay hybrid bionanocomposites. Poly Compos 16
40. Subasinghe A, Das R, Bhattacharyya D (2016) Parametric analysis of flammability performance of polypropylene/kenaf composites. J Mater Sci 51:2101–2111
41. Bocz K, Szolnoki B, Marosi A, Tabi T, Wladyka-Przybylak M, Marosi G (2014) Flax fibre reinforced PLA/TPS biocomposites flame retarded with multifunctional additive system. Polym Degrad Stab 106:63–73
42. Jeencham R, Supparkarn N, Jarukumjorn K (2014) Effect of flame retardants on flame retardant, mechanical, and thermal properties of sisal fiber/polypropylene composites. Compos B 56:249–253
43. Szolnoki B, Bocz K, Soti P, Bodzay B, Zimonyi E, Toldy A, Morlin B, Bujnowicz K, Wladyka-Przybylak M, Marosi G (2015) Development of natural fibre reinforced flame retarded epoxy resin composites. Polym Degrad Stab 119:68–76
44. Sudhakara P, Kannan P, Obireddy K, Rajulu AV (2011) Flame retardant diglycidylphenylphosphate and diglycidyl ether of bisphenol-A resins containing Borassus fruit fiber composites. J Mater Sci 46:5176–5183
45. Sain M, Park S, Suhara F, Law S (2004) Flame retardant and mechanical properties of natural fibre-PP composites containing magnesium hydroxide. Polym Degrad Stab 83:363–367
46. Shumao L, Jie R, Hua Y, Tao Y, Weizhong Y (2010) Influence of ammonium polyphosphate on the flame retardancy and mechanical properties of ramie fiber-reinforced poly(lactic acid) biocomposites. Polym Int 59:242–248
47. Schartel B, Braun U, Schwarz U, Reinemann S (2003) Fire retardancy of polypropylene/flax blends. Polymer 44:6241–6250
48. Nie S, Liu X, Dai G, Yuan S, Cai F, Li B, Hu Y (2012) Investigation on flame retardancy and thermal degradation of flame retardant poly(butylene succinate)/ bamboo fiber biocomposites. J Appl Polym Sci 125:485–489
49. Cogen J, Lin T, Lyon R (2009) Correlations between pyrolysis combustion flow calorimetry and conventional flammability tests with halogen-free flame retardant polyolefin compounds. Fire Mater 33:33–50
50. Suardana N, Ku M, Lim J (2011) Effects of diammonium phosphate on the flammability and mechanical properties of bio-composites. Mater Des 32:1990–1999

51. Manfredi L, Rodriguez E, Wladyka-Przybylak M, Vazquez A (2006) Thermal degradation and fire resistance of unsaturated polyester, modified acrylic resins and their composites with natural fibres. Polym Degrad Stab 91:255–261

52. Shukor F, Hassan A, Islam M, Mokhtar M, Hasan M (2014) Effect of ammonium polyphosphate on flame retardancy, thermal stability and mechanical properties of alkali treated kenaf fiber filled PLA biocomposites. Mater Des 54:425–429

53. Lilholt H, Lawther J (2000) Natural organic fibres. In: Kelly A, Zweben C (eds) Comprehensive composite materials, vol 6. Elsevier Science, pp 303–325

54. Alvarez V, Rodriguez E, Vazquez A (2006) Thermal degradation and decomposition of jute/vinylester composites. J Therm Anal Calorim 85:383–389

55. De Chirico A, Armanini M, Chini P, Cioccolo G, Provasoli F, Audisio G (2003) Flame retardants for polypropylene based on lignin. Polym Degrad Stab 79:139–145

56. Ferry L, Dorez G, Taguet A, Otazaghine B, Lopez-Cuesta J-M (2015) Chemical modification of lignin by phosphorus molecules to improve the fire behavior of polybutylene succinate. Polym Degrad Stab 113:135–143

57. Zhang J, Fleury E, Chen Y, Brook M (2015) Flame retardant lignin-based silicone composites. RSC Advances 5:103907–103914

58. Xing W, Yuan H, Zhang P, Yang H, Song L, Hu Y (2013) Functionalized lignin for halogen-free flame retardant rigid polyurethane foam: preparation, thermal stability, fire performance and mechanical properties. J Polym Res 20:234–245

59. Chen F, Dai H, Dong X, Yang J, Zhong M (2011) Physical properties of lignin-based polypropylene blends. Polym Compos 32:1019–1025

60. Dorez G, Otazaghine B, Taguet A, Ferry L, Lopez-Cuesta J-M (2014) Use of Py-GC/MS and PCFC to characterize the surface modification of flax fibres. J Anal Appl Pyrol 105:122–130

61. El-Sabbagh A, Steuernagel L, Ziegmann G, Meiners D, Toepfer O (2014) Processing parameters and characterization of flax fibre reinforced engineering plastic composites with flame retardant fillers. Compos B 62:12–18

62. Hapuarachchi T, Ren G, Fan M, Hogg P, Peijs T (2007) Fire retardancy of natural fibre reinforced sheet moulding compound. Appl Compos Mater 14:251–264

63. Hapuarachchi T, Peijs T (2010) Multiwalled carbon nanotubes and sepiolite nanoclays as flame retardants for polylactide and its natural fibre reinforced composites. Compos A 41:954–963

64. Rupper P, Gaan S, Salimova V, Heuberger M (2010) Characterization of chars obtained from cellulose treated with phosphoramidate flame retardants. J Anal Appl Pyrol 87:93–98

65. Subasinghe A, Bhattacharyya D (2014) Performance of different intumescent ammonium polyphosphate flame retardants in PP/kenaf fibre composites. Compos A 65:91–99

66. Le Bras M, Duquesne S, Fois M, Grisel M, Poutch F (2005) Intumescent polypropylene/flax blends: a preliminary study. Polym Degrad Stab 88:80–84

67. Dorez G, Taguet A, Ferry L, Lopez-Cuesta J-M (2014) Phosphorous compounds as flame retardants for polybutylene succinate/flax biocomposite: additive versus reactive route. Polym Degrad Stab 102:152–159

68. Yu T, Jiang N, Li Y (2014) Functionalized multi-walled carbon nanotube for improving the flame retardancy of ramie/poly(lacic acid) composite. Compos Sci Technol 104:26–33

69. Reti C, Casetta M, Duquesne S, Bourbigot S, Delobel R (2008) Flammability properties of intumescent PLA including starch and lignin. Polym Adv Technol 19:628–635

70. Nie S, Liu X, Wu K, Dai G, Hu Y (2013) Intumescent flame retardation of polypropylene/bamboo fiber semi-biocomposites. J Therm Anal Calorim 111:425–430

71. Li X, Tabil S, Panigrahi S (2007) Chemical treatments of natural fibers for use in natural fiber-reinforced composites: a review. J Polym Environ 15:25–33

72. Biswal M, Mohanty S, Nayak S (2012) Thermal stability and flammability of banana-fiber-reinforced polypropylene nanocomposites. J Appl Polym Sci 125:432–443

73. Zhang Z, Zhang J, Lu B-X, Xin Z, Kang C, Kim J (2012) Effect of flame retardants on mechanical properties, flammability and foamability of PP/wood-fiber composites. Compos B 43:150–158

74. Gallo E, Schartel B, Acierno D, Cimino F, Russo P (2013) Tailoring the flame retardant and mechanical performances of natural fiber-reinforced biopolymer by multi-component laminate. Compos B 44:112–119

75. Kandare E, Luangtriratana P, Kandola B (2014) Fire reaction properties of flax/epoxy laminates and their balsa-core sandwich composites with or without fire protection. Compos B 56:602–610

76. Srinivasan V, Boopathy S, Sangeetha D, Ramnath B (2014) Evaluation of mechanical and thermal properties of banan-flax based natural fibre composite. Mater Des 60:620–627

77. Rana A, Singha A (2014) Studies on the performance of polyester composites reinforced with functionalized Grewia optiva short fibers. Adv Polym Technol 33:21433–21443

78. Nair K, Thomas S, Groeninckx G (2001) Thermal and dynamic mechanical analysis of polystyrene composites reinforced with short sisal fibres. Compos Sci Technol 61:2519–2529

79. Matkó S, Toldy A, Keszei S, Anna P, Bertalan G, Marosi G (2005) Flame retardancy of biodegradable polymers and biocomposites. Polym Degrad Stab 88:138–145

80. Jang JY, Jeong TK, Oh HJ, Youn JR, Song YS (2012) Thermal stability and flammability of coconut fiber reinforced poly(lactic acid) composites. Compos Eng 43:2434–2438

81. Alongi J, Carosio F, Malucelli G (2014) Current emerging techniques to impart flame retardancy to fabrics: an overview. Polym Degrad Stab 106:138–149

82. Fang F, Xiao D, Zhang X, Meng Y, Cheng C, Bao C, Ding X, Cao H, Tian X (2015) Construction of intumescent flame retardant and antimicrobial coating on cotton fabric via layer-by-layer assembly technology. Surf Coat Technol 276:726–734

83. Fang F, Chen X, Zhang X, Cheng C, Xiao D, Meng Y, Ding X, Zhang H, Tian X (2016) Environmentally friendly assembly multilayer coating for flame retardant and antimicrobial cotton fabric. Prog Org Coat 90:258–266

84. Carosio F, Negrell-Guirao C, Di Blasio A, Alongi J, David G, Camino G (2015) Tunable thermal and flame response of phosphonated oligoallylamines layer by layer assemblies on cotton. Carbohyd Polym 115:752–759

85. Pan H, Song L, Hu Y, Liew KM (2015) An eco-friendly way to improve flame retardancy of cotton fabrics: layer-by-layer assembly of semi-biobased substance. Energy Procedia 75:174–179

86. Wang L, Zhang T, Yan H, Peng M, Fang Z (2013) Modification of ramie fabric with a metal-ion-doped flame-retardant coating. J Appl Polym Sci 129:2986–2997

87. Lin Z, Renneckar S, Hindman D (2008) Nanocomposite-based lignocellulosic fibers 1. Thermal stability of modified fibers with clay-polyelectrolyte multilayers. Cellulose 15:333–346

88. Dorez G, Otazaghine B, Taguet A, Ferry L, Lopez-Cuesta J-M (2014) Improvement of the fire behavior of poly(1,4-butanediol succinate)/flax biocomposites by fiber surface modification with phosphorus compounds: molecular versus macromolecular strategy. Polym Int 63:1665–1673

89. Zhou L, Ju Y, Liao F, Yang Y, Wang X (2016) Improve the mechanical property and flame retardant efficiency of the composites of poly(lactic acid) and resorcinol di(phenyl phosphate) (RDP) with ZnO-coated kenaf. Fire Mater 40:129–140

90. Misra R, Kumar S, Sandeep K, Misra A (2008) Some experimental and theoretical investigations on fire retardant coir/epoxy micro-composites. J Thermoplast Compos Mater 21:71–101

Conclusion

In order to reduce our environmental footprint, polymer industry has started to develop new materials based on natural resources.

Two kinds of biobased polymers can be listed. The first one corresponds to macromolecular structures already existing in nature as cellulose, lignin, starch, alginate and so-on. These structures are rich in oxygen, release relatively little heat during burning and are often able to char. Nevertheless, their thermal stability is limited. Other biobased polymers are made up of molecules synthetized from natural resources. Chapter 1 deals with this kind of polymers which can be processed using classical tools devoted to thermoplastics (as twin-screw extrusion and molding injection). These polymers as PLA (polyester) or PA11 (polyamide) have often synthetic counterparts of similar structure. Therefore their fire behavior is close to that of these counterparts. To improve their flame behavior, researchers have reasonably applied the strategies already available for oil-based polymers. Performances of flame retardant systems are not qualitatively different from those observed for oil-based polymers.

Not only polymers but also all additives used to modify their properties can be biobased to meet sustainable development. Intensive research is nowadays devoted to develop biobased flame retardants from various raw resources. Chapter 2 is an attempt to classify these new flame retardants according to the original bioresource. Biobased FRs can be used directly as they are, alone or as a component of a more complex system. This is especially true when the molecules are phosphorus-rich (as DNA or phytic acid for example) or charring (as lignin). In other cases, the raw molecules can be functionalized to become FRs. Most generally, functionalization involves phosphorylation. The research for effective biobased FRs just beginning.

Natural fibers can be used as reinforcement for composites. While they replace inorganic glass fibers, they are believed to change and degrade drastically the fire behavior of composites. In fact the review detailed in Chap. 3 show that the reality is more complex. Indeed, natural fibers can produce char themselves or in presence of charring promoter as phosphorus-based FR. In that case, natural fibers can be considered as a part of a multicomponent FR system. Fibers can be flame retarded through chemical modification. This strategy could lead to reduce the amount of flame retardants incorporated into the matrix, in order to avoid a drop of mechanical

© The Author(s) 2018
R. Sonnier et al., *Towards Bio-based Flame Retardant Polymers*,
Biobased Polymers, DOI 10.1007/978-3-319-67083-6

properties or processing issues. Nevertheless, up to now, the flame retardancy of biocomposites is not as good as possible when only reinforcement is flame retarded.

All the efforts reviewed in this book show that a major objective of researchers is to develop 100% biobased materials suitable for applications requiring high flame retardancy level. Nevertheless, "biobased" does not mean systematically "ecofriendly".

Printed in the United States
by Bookmasters

Printed in the United States
By Bookmasters